D0428505

Special Characters

DEY ST.
An Imprint of WILLIAM MORROW

Special Characters

MY ADVENTURES WITH TECH'S TITANS AND MISFITS

Laurie Segall

DEY ST.

HarperCollins books may be purchased for educational, business, or sales promotional use. For information, please email the Special Markets Department at SPsales@harpercollins.com.

FIRST EDITION

Designed by Paula Russell Szafranski
Title page spread image © BG Plus2 / Shutterstock.com

Library of Congress Cataloging-in-Publication Data

Names: Segall, Laurie, author.
Title: Special characters : my adventures with tech's titans and misfits / Laurie Segall.
Description: First edition. | New York : Dey Street, 2022.
Identifiers: LCCN 2021034426 (print) | LCCN 2021034427 (ebook) | ISBN 9780063016446 (hardback) | ISBN 9780063016460 (ebook)
Subjects: LCSH: Segall, Laurie. | Journalists—United States—Biography. | Women journalists—United States—Biography. | Technology—Press coverage.
Classification: LCC PN4874.S44 A3 2022 (print) | LCC PN4874.S44 (ebook) | DDC 070.92 [B]—dc23/eng/20211103
LC record available at https://lccn.loc.gov/2021034426
LC ebook record available at https://lccn.loc.gov/2021034427

ISBN 978-0-06-301644-6

22 23 24 25 26 LSC 10 9 8 7 6 5 4 3 2 1

To all those who live their lives

full of lobster moments

A Note from the Author

The events and experiences detailed on the pages you're about to read are all true and have been written as I remembered them, to the best of my ability. Some names and identifying details have been changed to protect the privacy of the subjects I wrote about. I've also condensed and combined stories to fit a wild decade into 368 pages and re-created dialogue to best represent that time period. Though conversations come from my recollection, my brain is not yet powered by artificial intelligence, so they aren't written to represent word-for-word documentation; they come from dozens of reporter notebooks, personal journals, text messages, and notes typed up in my phone that thankfully never disappeared. Luckily, I'm a bit sentimental and have kept detailed journals throughout my career. Also, most of my important conversations with tech titans were recorded live for everyone to see. All of the dialogue and stories you're about to read are retold in a way meant to evoke the real feeling and meaning of what was said at the time, in keeping with the mood and spirit of the event. Happy reading.

Contents

Special Characters

Prologue

THE LOBSTER

A wise-looking rabbi named Dr. Abraham Twerski gazes into the camera, his long white beard giving him an extra air of importance against the black backdrop as he explains the anatomy of a lobster.

The rabbi begins his online sermon with a description: a soft mushy animal living inside a hard shell. At first it's comfortable there, but that rigid shell can't expand. The lobster feels pressure and discomfort as it grows larger, and the shell becomes tight and confining. Eventually, he explains, the lobster hides somewhere safe while it casts off the old shell and grows a new one.

It's important to note, he continues, that the lobster is incredibly vulnerable in this moment of growth, as it hides under a rock formation, protecting itself from predatory fish and other creatures of the sea. But the discomfort is well worth it. Stress may begin the process, but by the end, the lobster has a new shell that fits.

Of course, the lobster continues to grow, and eventually the new shell becomes confining, and the lobster must repeat the process.

The only way for the lobster to grow is to feel the discomfort, stress, and vulnerability that come along with shedding its shell, to grow a new one. Over and over again . . .

CHAPTER 1

Showtime

"Ma'am, are you okay?"

The cop had pulled over and his partner rolled down the window.

Maybe it was my heels. I must have been wobbling as I walked across Second Avenue. At five-five, I'm committed to wearing at least four inches, but that doesn't mean I do it gracefully.

The streets, which were usually filled with herds of NYU students bouncing between East Village bars, were now empty.

I glanced at my iPhone as the driver of the police car cut the engine.

It was 4:03 A.M. on August 18, 2008. My twenty-third birthday.

I adjusted my white-collared button-up that I'd deemed appropriate attire for anything professional and gave my best smile to the baby-faced cop and his older partner.

"Oh yeah, I'm totally fine!"

I was far too enthusiastic for this time of day. My body vibrated with the kind of energy you have before the years tick by and the circles under your eyes deserve their own nicknames. The kind of enthusiasm the policemen would surely deem a cover-up for a twenty-three-year-old trying to appear sober.

"It's a bit early to be walking here," he said. "This isn't a safe area for a woman alone, this time of the morning. Do you need a ride?"

I wasn't sure how true that was. The East Village seemed like a pretty safe place, even at 4:00 A.M. A part of me doubted whether I would've found myself in this same scenario had I not been a young white woman, making her way to work. I hesitated. Should I ride in the back of a cop car or wait for the subway?

After a bit of mental calculus, balancing the subway or a free ride, I climbed into the back seat.

"Fifty-Eighth and Eighth," I called.

"So, where're you headed this early?" the younger cop said, looking in the rearview mirror.

"It's my first day on the job." I paused. "At CNN." It was the first time I'd said it. I let the words linger in the air.

"Fancy!" the older cop barked, winking, then taking a swig of coffee.

I grinned. I didn't tell him it was a tentative position as a freelance news assistant. And I certainly didn't tell him that my biggest accomplishment so far as a budding journalist had been getting a toe in the door. It involved every skill I had—tenacity without being annoying, creativity, and above all, scrappiness. I'd been working toward this moment for a decade.

As the police car came to a stop, I thanked the officers and exited to a lightening sky. The sun had yet to rise on the postcard view of Central Park; the streetlights danced across the buildings. It wouldn't be long until the scent of vendors selling honey-roasted nuts and coffee for morning commuters filled the air. I took in a deep breath and looked around at the quiet streets, listening to the yellow taxis cruise by. As I walked through the revolving doors of

1 Time Warner Center, I felt like I was walking onto a movie set, complete with opening credits to my own life.

Inside the building, everything was dark and marble, and in the early hours, completely silent, apart from the clicking heels of rushed producers making coffee runs before the morning shows began, entering and exiting seamlessly with electronic badges. The security guards sat in an authoritative row near the doors and one of them noticed me stalling near the turnstile.

"Miss?" he said, looking over, his eyebrow raised, his demeanor questioning how I'd made it into the building.

"It's my first day!" I said with unchecked enthusiasm.

He grunted and pointed me to another guard, who ushered me to the seventh floor. There, yet another guard who didn't share my enthusiasm took a snapshot of me smiling too widely, adjusting my collar in time for the flash.

Then I was shuttled back down the elevator to the fourth floor, where dozens of desks dotted the open newsroom, cameras stretching their necks like giraffes hovering over the jungle of tangled monitors and blinking screens. Day or night, the newsroom was a labyrinth of incoming feeds, lights from the control rooms flashing on and off. Edit bays, where TV segments were cut to be later aired on shows like *Anderson Cooper 360°*, lined the hallways.

I'd spent the last week researching the producers I'd be working for, the ones who'd reported on 9/11. They were the hard-nosed journalists who ran toward the buildings, risking their lives to relay the news, documenting the horror of that day almost seven years ago. I had scrolled through YouTube videos, listening to their voices stating the facts. I studied their names, pressing *pause* and *play* over and over again. They were the real deal, and I was entering their bullpen. I'd be training under the best.

I took a seat and waited with a handful of other freshly minted news assistants. Despite my excitement, I knew where we stood. We were freelancers, entry-level nobodies with a one-year trial period to impress the right people, or else get out. I knew I'd have to find a way to make myself stand out in a sea of ambitious journalism hopefuls who'd made it this far. With relentless work, the right advocates, and perfect timing, I'd have a shot at a full-time position.

As the others exchanged tips and small talk, I sat silently, watching as producers stalked in and out of the newsroom, reporters marched around in their heels, and desk managers barked orders into their phones. I felt both terrified and like the luckiest person in the world.

CHAPTER 2

Wrecking Ball

My path to the newsroom started in middle school, the awkward and uncomfortable years when we have yet to settle into ourselves—the years we're told we'll grow out of. I was a pudgy preteen in an oversize flannel sweatshirt draped over khaki pants—the ones big enough to hide my discomfort in my own body, and the insecurity that came along with that.

It was a bright autumn morning when my mother sat my brother and me down on the pinstripe couch and told us that our father didn't love her anymore. She closed the curtains, and a new era began. It was painful, and I struggled to adjust to the unsteady new family dynamic, one where my father was no longer an everyday part of my life and my mother was grappling with a broken heart and the anger that now filled our home. Our gray wooden house at the top of a long driveway in the leafy suburbs of Atlanta had a shadow cast on it. I retreated into my journals, where I wrote about the battles happening under our roof, stories about strangers, and suburban observations. I poured myself into the act of writing, filling worn notebooks with song lyrics that resonated with the isolation I felt. I held tightly to my friends' families, to their warmly lit homes full of laughter and home-cooked meals.

In the meantime, my mother tried to comfort us by taking us to a suburban mecca: McDonald's.

We went at least once a week, my mother trying to help us feel normal, to give us something to look forward to. During this time, McDonald's introduced a new promotion: if you supersized your order, you could play Monopoly. Dutifully, we supersized our fries and sodas, peeling off the thumbnail-size playing pieces, hoping to win a prize or get a monopoly. As in the real game of Monopoly, everyone wanted Boardwalk. It was the most expensive property, and if you got that royal-blue tab, you could win a million dollars.

Within a year, I'd gained fifteen pounds and was one card away from a million dollars. All I needed was Boardwalk. The quest became my escape from reality—from the snobby suburbs of Atlanta, the divorce, the pain, and the overwhelming desire to fit into a world that seemed carved out for other people.

Perhaps if we had a million dollars, my mother wouldn't worry about our shifting financial reality. If we were to achieve Boardwalk, the cruel middle schoolers on the bus, whose mothers refused to accept mine into their stifling circles, whose worst weapons weren't their sharp words but their general disinterest, would pay attention to me. If we won the game, maybe my father would call more frequently.

I was a "good girl," not by choice, but by instinct. Instead of staying out past curfew and dabbling in mischief, I felt a responsibility to be my mother's "plus one," to stay by her side and take on an adult role that included emotional support. My father, despite being a well-loved doctor, bristled at my anger. His new home was on the other side of town, but it felt like the other side of the country. Our family collectively ached, all of us unable to fully process the pain, manifesting in choose-your-own-adventure coping mechanisms, most of them unhealthy: Class photos with tack holes stabbed through the eyes. Doors slammed, and fingers and hearts broken. Our house, once filled with nightly piano recitals and spaghetti and meatball dinners, was paralyzed with anger.

I remembered my parents touching only once, years before. I'd held it with me, playing it over in my head like a song on repeat. My mother and father were driving me and a friend to the movie theater. We sat in the back of our white minivan, chattering as the car filled with music. Frank Sinatra's "The Way You Look Tonight" began playing. My father looked over at my mother, his mouth creeping into a warm smile, and placed his arm around her shoulders. When she beamed back at him, I knew this was their song.

Now the music was replaced with muffled rage. As chaos swirled around me, I felt invisible. But I always held on to the hope of something better—of Boardwalk. Something better was just one meal away. Somewhere around the corner "spectacular" existed. A life where tight-knit communities in the beautiful green suburbs were nonjudgmental of broken families, where holiday dinners had five types of pie and heaps of laughter, where I wasn't numb, and scared, and invisible—where anything was possible.

I cried myself to sleep when my brother went to boarding school, far from our chaotic home. I dug half-moons into my palms when I transferred to a new school in hopes of branching out from Holy Innocents' Episcopal School, a southern, conservative high school where I was one of the only Jewish girls. No matter how much I had tried to fit in, I stuck out; I didn't have blond hair or blue eyes or parents who owned part of the Coca-Cola empire. When one of the boys told me I was going to hell because I didn't believe in Jesus, I laughed reflexively and shrugged, wondering if there would ever be a place where I didn't have to try so hard to laugh off the feeling of being an "other." I was uncomfortable in my own skin, in my tiger-striped retainer and frizzy hair that wouldn't stay behind my ears.

But transferring to a new school, I felt even more invisible than I had before. The "nice Jewish girls" my mother told me about avoided eye contact. Our imperfect family wasn't warmly welcomed in those circles.

Within a year, I transferred back to my old school, this time making my own space. Instead of attending tailgate parties, I listened to ska bands and drove myself to the Roxy Theatre in Buckhead, a short drive from my home. There, I'd watch string-cheese-thin musicians drape themselves over guitars and sing stories about another world. CDs by Reel Big Fish, RX Bandits, and Something Corporate were strewn about my used white Ford Explorer. I'd blast "Punk Rock Princess," windows down. In reality, I was far from punk rock, and even further from a princess.

At sixteen I became managing editor of the student newspaper. I had opinions on everything and penned op-eds on why gay people should marry and a particularly blistering piece on the insensitivity of the county fair's bouncy slide, which was in the shape of the *Titanic*. But I spent most of my time writing a column called "Spotlight," where I had the opportunity to profile whoever I wanted. Being one of the least athletic students no matter the sport, I came to know the sidelines. So instead of writing profiles of the sports stars, I interviewed those behind the scenes.

Coach Red, the track coach who was well into his eighties, was tall and had kind eyes. It had been many decades since he'd run a lap and the other kids laughed when he tried directing sprints. During our interview, we sat atop the mats in the corner of the gym, and I asked him questions that I'd scrawled out in my notebook. He lit up when he discussed meeting his wife decades before and fighting in World War II. It was the first time I experienced the "light bulb moment" in an interview: that moment when the curtain falls and reveals something about a person that changes your point of view. As we were nearing the end of our interview, he opened up about his struggle with Parkinson's disease. Under the lights of the basketball court, he started crying. He talked about living the rest of his life knowing what it would be like to die soon. The man people snickered at as he moved slowly, directing runners like a sleepy cop at a traffic stop, was gone. In front of me was an inspiring human who'd fought a war, found love, and was now beginning a new battle, staring at his mortality. The conversation was raw, it was real. His words were the most meaningful thing anyone had ever said to me.

So what if I can't run a lap? I want to tell these kinds of stories for the rest of my life. And so, for the last two years of high school, I used my newspaper real estate to profile other castoffs, including Annabelle, the ancient library lady who definitely had biblical secrets, and my own grandpa, also a war hero, who was one of the first doctors to desegregate waiting rooms in the South. I dubbed these profiles "corner stories," and through them, I found a sense of belonging, a solace in other people's inability to fit into the norm.

After graduation, I couldn't wait to reboot my life at the University of Michigan, in a prestigious program for wannabe writers like myself. Most people from my high school stayed nearby, attending schools like Auburn, the University of Georgia, or Clemson. Yet despite the ongoing warnings about the frigid weather and gray skies, I knew one thing: I wanted distance. Every single college I had applied to was hundreds of miles away.

The summer before matriculating, I, like so many others, created my first Facebook profile. The company had just opened up its platform beyond the Ivy League schools to allow other college students to join, billing itself as an "online directory" connecting people through social networks. Signing on to Facebook—which had recently dropped the "the" from "The Facebook"—made me feel like a pioneer in a new digital era.

"I think you just add me?" my friend Jackie said excitedly as we huddled over her computer, duffle bags packed for our big exits.

We both now had a Facebook friend, each other, and we promised to stay connected as we went our separate ways.

After landing in Ann Arbor, my mother drove me to campus, where I found my dorm room and dropped my suitcase on a well-worn mattress. Before leaving, she wrapped her arms around me for a beat too long. "I love you, honey," she said, before slipping out the door. For so long it had been the two of us, but today I was entering a new phase of adulting. Shrieks and laughter filled the hallway. Unbeknownst to me, the prestigious writing program I'd been accepted into had the college's largest dorm rooms, a secret known to a subset of Long Islanders who'd applied to the program for its larger living quarters. Day one in Lloyd Hall, it was me and all the Long Island girls—some with big goals, all with big rooms. I was both thrilled and shocked. Here I went from being one of the only Jewish girls to looking similar to every other girl around me.

But even though I looked the part, I still didn't fit in. The stick-straight hair, Ugg-boot, Hard Tail–legging uniform was a religion I had yet to learn. Gone were the shaggy-haired, door-holding boys from the South. The guys at U of M had spiky, gelled hair and came from wealthy, fast-talking enclaves. I'd entered a new world.

Desperate to find my place, I joined a sorority because, according to people who seemed to know more than I did, that was what out-of-state students did. But while I looked like the other girls in Kappa Alpha Theta, on the inside, I felt like the vintage E.T. doll I'd bought at a thrift shop in Ann Arbor. My stint at the sorority ended after I was scolded for not cheering loudly enough during a recruitment event. Looking around at my "sisters" clapping in unison, chanting "I just can't get enough of Theta. Go, Theta!" I realized I'd reached my limit and resolved to find a home beyond Washtenaw Avenue.

During this time, I never stopped journaling. My habit of filling notebooks carried over to coffee shops in Ann Arbor. I found new corner stories and characters, like the elderly man with a long white beard who played harmonica in The Diag on campus. I dropped my green JanSport backpack next to him and listened to his music, channeling the story of a man who many thought was homeless but was actually a retired professor struggling to reconnect with his son.

Journaling was my main outlet for synthesizing my thoughts and feelings and developing an understanding of the world around me. It was my safe place, where I could write about whatever I wanted—not the five-paragraph essays one of my strict English teachers in high school had demanded. Other than my experience writing for Mrs. Klepper, who quickly became an advocate for me on the newspaper staff, writing in school was always a chore, not a channel for storytelling or creativity.

But in my writing programs in college, everything changed. An instructor introduced me to the work of Joan Didion, whose writing thrilled me. Didion's sentences were long and descriptive. She captured people and ideas, and applied meaning to the everyday events that to many might seem mundane. I dog-eared pages of her books and essays: a bride in Vegas became a telling anecdote, loaded with meaning. There was something unique about Didion's voice and the way she universalized her feelings. I was drawn to her honesty. I'd never been encouraged to write in such a conversational way. Didion was raw and intense. She was everything I feared: She didn't play by the rules. She made her own. The writing program I was part of became an outlet for creativity, and a place where the flurries of sentences and observations I'd spent years piling into journals found a home. The more I became engulfed in stories, the more I knew I wanted to make them my career. I doubled down on any writing opportunity I could find and began volunteering at the college television station, writing segments for WOLV TV.

My junior year, I got my big break. I was introduced to an editor from *Glamour* through a college friend whose cousin was connected to the magazine. The editor offered me the opportunity of a lifetime: to go undercover to a "purity ball," a ceremony where young girls sign purity pledges to their fathers, who become the "safekeepers" of their daughters' virginity until it can be "given" to their husbands—all happening in my home state of Georgia.

I was thrilled. This was the ultimate corner story, and I was going undercover—with my dad. At the time, my father and I were barely speaking. Weeks before, he'd informed me that he was getting married to Harriet, his girlfriend, with whom I had no relationship. I swallowed my shock, choked out congratulations, and blurted out my question: Would he be willing to attend a purity ball with me?

Over the years, our relationship had gone from rocky to tumultuous, but he was genuinely happy to take part in building my career. It was my first foray into journalism beyond academia. And if I did well, it would open doors.

I returned to Atlanta and together we flew to Jekyll Island, Georgia, which, according to Google, is known for golf, faith, and sea turtles. I pushed away the discomfort that accompanied the only one-on-one time my father and I had had in over a decade and focused on the assignment. The idea behind the purity ball was fascinating . . . and weird. Girls as young as nine were signing pledges, declaring their fathers the safekeepers of their sexuality, with a backdrop of ballroom dancing and motivational speaking. At its core, the event was meant to highlight the importance of father-daughter relationships. *How ironic,* I thought, as I offered my dad a packet of airplane peanuts.

Within two hours, we were walking into the lobby of the convention center. My heart stumbled in my chest. We were surrounded by preteens in sparkling princess ball gowns, their hair wrapped in braids and crowned with tiaras. I tugged at my snug white dress. Damn, it was too formfitting. The fact that I even *had* a form meant I was one of the oldest attendees by far.

Dad adjusted his glasses and stood uneasily in the check-in line behind a cluster of girls with long white gloves stretched to their elbows, clutching their fathers' hands. I crossed my arms and held my breath as a man asked for our last name.

"Segall," my father said.

I clenched my teeth, convinced that the name alone would give me away—as it had in the many debutante balls, cotillion classes, and chapel events I'd attended in my youth.

The man found our name on the list.

"Head on in, sir," he said with a thick southern accent, pointing my father to the open doors of the ballroom.

Level one. Pass, I told myself.

Dad and I moved past the ice sculptures and colored balloons, weaving through groups of girls in puffy dresses to our table, which was labeled with a large white card that read, "Gentleness." At each of our seats, there was purity paperwork to fill out.

Dad reviewed the card in front of him—a father's purity prayer that asked him to be a positive role model (*fair request*) and to protect my purity (*hmm*). I reviewed mine. Remain pure in all areas (*I can't swear to that*). The pledge declared my dad the high priest of my sexuality.

Dad was a good sport. I watched, amused, as he was asked to join the "fathers circle" to recite the pledge prayer with the other men. A distant, analytical thinker, he stood and repeated a passage vowing to be my sole male relationship and the protector of my purity until marriage.

Then, standing with the girls, I dedicated my virginity and purity to my father until marriage. As I watched the girls reciting their pledges, I sensed the gravity of the promise, but there was no way that paperwork I would have signed as a ten-year-old could represent the twenty-one-year-old me. Especially when it pertained to something as personal and evolving as my sexuality.

I couldn't wait to write about it. When I returned to the University of Michigan, I composed a long piece about the implications of young women signing away their sexuality before even knowing what it was. It was thoughtful and nuanced—my first think piece. When I sent it to the editor at *Glamour,* I truly believed this was my big break.

And so began my worst nightmare.

It was a typically gray Ann Arbor afternoon when I walked into Amer's Deli and my flip phone rang. It was the editor from *Glamour.* She cut straight to it:

"Laurie, quick question. Are you a virgin?"

I racked my brain as I ducked behind a shelf of Utz potato chips. While it might seem like I had nothing in common with a pack of purity-ball pledgers, I, too, was a virgin. But nowhere in my essay did I happen to mention that I'd never had a boyfriend, or that I had trouble with basic male interaction. A part of me still felt like the awkward middle schooler searching for a seat on the bus. I'd had a slow start, and then struggled to catch up. It seemed like I was the only one left who hadn't had the full sexual experience, but the more I watched the gelled-haired guys at U of M, the less rushed I felt to experience it. So I'd resigned myself to waiting until I found the right person.

"I mean, it's not like I'm waiting because of a pledge," I began rambling to my editor, who I imagined was perched in a gleaming office with soaring views of Central Park. "I just didn't fit in at my high school, and the guys in my dorm spend more time on their hair than I do . . . It's too late to lose my virginity casually, and besides, there are no viable options."

The editor wasn't listening. "We want this in first person," she cut in. "More personal."

I must have said something before I hung up. If I rewrote this piece the way the editor wanted, I'd *never* find a boyfriend.

I failed the test. No number of drafts could make me declare my virginity to the University of Michigan or, well, the entire country. But the piece wasn't killed. Every draft returned felt less and less me, more simplified, less nuanced. The "think" aspect of the piece was slowly replaced with lines written by an increasingly frustrated editor. I knew I was inexperienced, so I picked my battles, afraid to speak up when asked to write certain sentences that I felt cheapened the event.

An early proof of the magazine arrived in the mail, and I tore through it to find my story. There was my byline, followed by a short piece written by someone else. In the first paragraph, it declared "At 21, I'm still a virgin" and went downhill from there, explaining how I'd "faced the regret of a meaningless hookup," which, in all my awkward interactions with men, I hadn't. Above the essay was the picture of me in the white dress, beaming next to my father, holding my signed purity pledge.

There was no doubt that after this got out, I'd be a virgin forever. I knew it would be catnip for the Ugg-boot brigade at Kappa Alpha Theta.

The timing of my nationwide humiliation worked out well. Thankfully, I had signed up for a study-abroad program in London and left the country just as the magazine hit the newsstands. Jennifer Connelly's smiling face beamed from the cover, alongside a bold headline offering fifty shortcuts to a sexier body. Inside the dozens of editions lining the newsstands was my article, "I Crashed a Purity Ball."

My virginity was everywhere; the article, which was supposed to be limited to the American edition of the magazine, soon appeared internationally as well, haunting me from my London flat. At twenty-one, I had an International Declaration of Virginity under my belt, along with an important realization: I didn't want to work for a women's magazine. I needed to get as far away from this experience as possible. Forcing myself to breathe, I challenged myself to find a summer internship at a news outlet as opposite to *Glamour* as possible. While I had a year until I graduated, it was common knowledge that a solid internship your junior year could lead to a job opportunity by graduation day. If I was going to accomplish my goal of breaking into the news industry, I had to intern at a company I wanted to work for after college.

I needed to get a job at CNN.

Over the next few months, from my twin bed in Paddington, I mastered the art of the cold call. I tracked down the head of CNN's human resources and called her once, twice . . . many times. With each message, I toed the fine line between stalker and persistent. Somehow, I got the balance right and received a response.

My victory was quickly tempered by rejection. Lots of it. I was turned down for internships with Anderson Cooper's team (my dream job); Glenn Beck's show (in retrospect, maybe not the worst thing); and one or two other shows that no longer exist.

Eventually I landed at CNN International for a summer internship. There were two producers and a whip-smart correspondent with short hair and a warm smile. Her name was Maggie Lake, and all I wanted was to do great work for her.

On my first day as an intern, I walked around the cafeteria at lunchtime, tray in hand, contemplating where to sit. Since my unit was small, there were no other interns to befriend. The Anderson Cooper crew, on the other hand, immediately formed a clique. They had a certain aura, and in all fairness, if you were logging tape for the Silver Fox, you'd have pep in your step, too. The Anderson posse was led by a blond guy who looked like a Disney prince and wore a different patterned tie every day; he would later become an Instagram influencer. But this was years before Instagram even existed.

At a loss, I found a food station and stood in line to pay for an overcooked burger.

"How's it going?" a middle-aged man asked, looking in what seemed to be my direction. He was a large guy in a striped polo shirt and wire-framed glasses. It took me a second to realize he was speaking to me.

"It's my first day, so I'm just trying to get by," I said with a candor that surprised me. But something about him was comfortable and inviting. I immediately trusted him.

I scanned the room. The Anderson clique huddled at a table with no empty seats. Should I sit alone? Was that sad?

"You're in luck," the man exclaimed, his vivid eyes lighting up behind the glasses. "It's my first day, too! I'm Ross."

An ally! I let out a sigh of relief. He invited me to sit with a group, and as we got closer, I realized they were older and shared the kind of ease that only comes from decades of working together. As soon as I sat down, it was clear that I'd joined a group of old-timers, Ross included.

"Sorry about that," he said. "I'm not actually new here. I work in the edit bays, creating TV packages that air on prime time."

I released my anxiety in a burst of laughter. He invited me to stop by, so in between live hits for Maggie Lake, I spent my free time in his edit bay, watching as he pieced together stories like puzzles—sound bites and images that turned into television. He showed me how hours of tape, scattered shots of anchors around the world, and images became three-minute pieces that landed on *Anderson Cooper* or *Larry King Live*. Ross worked in high-pressure situations, weaving images, words, and sound to create nightly mini masterpieces.

One day led to another, and before long I was spending more and more time with Ross. I found comfort in his edit bay; my conversations with him were the kind I could have had with my father, asking for career advice and opening up about my insecurities.

When I wasn't in the edit bay with Ross, I was at on-site shoots assisting with interviews and providing backup for producers. Since Maggie's team was so small, I got to do more than just log interviews, like the Anderson Cooper interns did. This was long before the days of automated transcripts, where services transcribed recorded interviews. Instead, interns listened to hours of taped interviews and logged every word for producers, so they could piece together news packages. The process was excruciating. An hour-long interview meant spending half a day, depending on how quickly your fingers moved across the keyboard, typing out every word the interviewee said.

My assignments sent me across the city and all around the building, and I took the opportunity to speak to *everyone*: producers who typed hurriedly into their phones as they waited in line at Starbucks on the tenth floor; associate producers who searched for footage at the tape library; and my favorite security guard, Gary. I became a frequent visitor to the control rooms, watching the operators behind the scenes. I befriended Roger, a welcoming camera operator in Studio 52. We discovered a shared love of Disney movies and show tunes, and he soon became my accomplice for impromptu karaoke sessions in his soundproof control room, where he'd pull up *Little Mermaid* karaoke on YouTube and give me teleprompting tips.

Every day I learned a different shortcut that opened another door and unearthed a new secret. For example, I quickly figured out the power of befriending the photojournalists responsible for shooting Maggie's segments. Befriend the shooters, and the shoot would be much smoother. Interns who didn't respect a shooter's schedule and didn't take time to organize productions correctly were in for a rude awakening. Crossing the photojournalists, many of whom had been at the network for ages, made life much harder for the interns trying to organize a shoot down the line.

After three months of absorbing every detail I could during my short stint, the internship ended. I stopped by Ross's edit bay with a handwritten card, a hug, and one conclusion: I'd mop floors to stay in the newsroom.

Arriving back in Michigan after a semester in London and a summer in New York City, I wasn't prepared to return to the college bars and sorority-lined streets. I was ready to move on from football games and tailgates, from frat parties and my small bubble, and into something bigger.

The following spring, after I'd graduated, Ross made a call to CNN's breaking news desk and got me an interview with the managers of "the Desk." In the thirty-minute phone interview, I spoke with two intimidating, no-nonsense voices on the other end of the line—managers Gene and Eden—who asked me basic questions about my internship and news knowledge. I couldn't read the inflection in their voices as we ticked through the interview questions, and after what felt like an emotionless call, I hung up wondering whether our discussion had been wonderful, terrible, or somewhere in between. Miraculously, I received news a couple days later: it had gone well. I was in—kind of. My official title was "freelance news assistant," and the job description was to roll prompters for anchors, mic up guests, make calls to investigate breaking news stories, and help producers in a time crunch get news on-air quickly and create television packages for the shows. Hours varied: 6:00 A.M. to 2:00 P.M., 11:00 A.M. to 7:00 P.M., or 3:00 P.M. to 11:00 P.M.

With a few weeks' notice, I moved to a shared three-bedroom apartment in a five-floor walk-up in the East Village. My neighbor Mario, who dressed in long gowns and sometimes called himself Maria, sang beautiful opera ballads at all hours. Directly underneath my bedroom, a local bar called Bua, run by Irish bartenders, offered drink specials to residents; it soon became a place to call home.

But under no circumstances could I get too comfortable. My job at CNN would last only one year before my freelance hours ran out. After that, the network couldn't legally keep me any longer without offering me a full-time position with health insurance. I had less than twelve months to charm someone into hiring me as a full-time employee.

CHAPTER 3

The Bullpen

My first days as a news assistant were a whirlwind of notetaking and survival. I was only a floor below Maggie's small unit, but I was far from the cocoon of my summer internship. Now I was in the bullpen, where assignment editors barked hard-to-grasp orders that I'd scribble into my reporter's notebook before fully understanding their meaning.

"Segall! We need you to ingest tape on seven for Berke, and then log the interview for seven P.M. Let us know when you're done. Tapes Desk."

Oh my god. Was that even English?

"Yes! Of course!" I'd say, grabbing a pen and praying that another, more senior assistant would help me translate.

The other assistants proved either helpful or harmful, with little in between. Within days, I'd identified the ones who were willing to help, and the ones who were happy to watch me fail miserably. The helpful ones recited control room numbers, gossiped about the producers, and relayed teleprompting-nightmare stories.

"I remember when the teleprompter crashed when I was prompting for Campbell Brown," a news assistant named Erin recalled, her eyes growing wide. "The whole control room stood up and screamed at me, 'PROMPTER!'"

"They didn't call you by your name?" I asked, slightly horrified.

"Oh no. When you teleprompt, you are nameless. And *never* correct them," she responded, referring to the producers who sat side by side in the control room orchestrating the show from what felt like a spaceship command center.

I nodded knowingly. The "prompter" position was low on the totem pole, but important. You were in charge of manually scrolling words on-screen, which the anchor read live on-air. Mess it up, and you made a room full of enemies, including the anchor. Maybe it wasn't a bad thing that we weren't referred to by name.

Erin laughed as she recalled "the incident," but it was clear she was still traumatized. "I'm pretty sure someone almost threw something at me," she said with forced cheer. "I nearly quit that day."

I had yet to learn how to teleprompt, but I was already having heart palpitations.

"That sounds like a nightmare."

"Oh, you haven't prompted for Sandra yet," she replied. "Just make sure you don't make eye contact."

"Seriously?"

"No joke."

Beware of Sandra, I scribbled into my notebook.

Day three in the bullpen, I met Deb.

"'Kids'?" I asked, referencing one of my favorite MGMT songs.

"Yep," she said, unimpressed by my ability to identify the music blaring from her headphones.

"Me too," I said, plopping my iPod in front of her.

She may have been disinterested, but it was 6:00 A.M., and we were the only news assistants on the shift. Within hours, other news assistants would arrive to join us in the bullpen, but for the time being, her options for human contact were slim.

As we awaited our assignments, we learned that we had both graduated from the University of Michigan and neither of us had felt like we belonged there. While I'd been a woefully miscast sorority sister—a southerner who'd landed in a sea of gelled hair and gossip—at five-ten, with long, bushy dark brown hair tamed by keratin treatments, Deb resembled a "nice Jewish girl" who'd fit seamlessly into the sorority parties I'd come to know (and fear) during my Lloyd Hall days. But she was different: she had no ability to "play the game." She was terrible at small talk and couldn't be bothered to have the types of conversations that occurred over jungle juice in sticky frat houses.

Deb grew up in Tenafly, New Jersey, in a suburb filled with gossip and expectations set by the Jewish mothers at the car pool, and had felt a similar sense of isolation at the University of Michigan.

"Segall, get the guest from security and bring them to Studio 53 for their segment," an assignment editor barked midday as the newsroom awoke to phones ringing like alarm bells and producers shouting updates over their computers.

"Okay!" I said enthusiastically. I looked around; most of the other news assistants had disappeared.

"That guest is a nightmare. He made Sarah cry once. Also . . . he's creepy," Deb whispered. "Next time he's on the schedule, you have to disappear around his arrival time, for self-preservation."

"Is that why everyone's missing?" I said, looking at the empty desks.

"Yep. Oh, and make sure Esther likes you," she said, referring to the bureau's business manager in charge of all the news assistants' schedules. "The more she likes you, the more hours she'll give you."

Deb quickly became my whisperer, looking out for me when no one else would. We were both extremely independent, convinced marriage would be a tough sell, and had a fear of societal norms: baby showers, registries where people sought expensive silverware, and growing up in general. We had an easier time talking to people struggling with loss than those who were defined by gains. We were both disorganized, ambitious, and empathetic. We were perfect allies for the newsroom, and even better candidates for best friends.

Over the next days, I started to notice her habit of tugging on a strand of her hair. In the background of no-nonsense news commentator Jack Cafferty's live shot, I could spot her on-screen, one of the people dotting the newsroom chairs, yanking at the same strand for hours, waiting for her next assignment.

After one of Cafferty's viewers called in to complain about "distracting people in the backdrop," one of the newsroom executives immediately called Deb into his office to lecture her. When she exited his office, I could see her blinking back tears. I caught her eye and gave her a reassuring smile, offering up as much news-assistant solidarity as a facial expression would allow.

"Should we grab a drink later to celebrate the most interesting thing to happen on Cafferty's show?" I whispered to her as she walked by. She gave me a grateful glance.

In reality, it was anxiety that kept her tugging on that same strand of hair. She didn't have to tell me—anxiety was like our secret handshake. I understood the feeling so well. Sometimes I watched people having conversations, feeling like I was in a movie, or was a ghost, watching from afar. I circled phrases in my head, wondering what to say, how to partake, what people were actually thinking. I overanalyzed every interaction, which was exhausting. Alcohol helped, until it did the opposite. I was nervous and restless, and so was Deb. But the moment we found each other, it felt like a deep exhalation.

Deb was one of the most thoughtful people I'd ever met, and I soon learned that she was also an incredible photographer. She took portraits of people—strangers on the subway, elderly women in parks, pedestrians on the street. We walked from the Time Warner Center down to Forty-Second Street, the flashing neon billboards blinking as throngs of tourists snapped photos with a life-size Batman and Minnie Mouse. Deb wanted to get a picture of Elmo, without the mask.

"Good luck," I said, as we approached. But within a couple of minutes, Deb, who spoke fluent Spanish, was deep in conversation with Elmo from Peru. He lifted his mask, revealing an elderly man with deep-set eyes and worn, leathery skin. She snapped a photo, and he immediately replaced his mask.

When Deb took photos, her subjects didn't look better or worse; they looked like their true selves. She captured everything about them—years of pain, a crossroads, an unseen love story. But as extraordinary as Deb was at pointing and shooting, she was that bad at faking it and schmoozing. Smiling at unreasonable bosses and reporters with hair-spray helmets, navigating tricky newsroom politics—those were my abilities. I'd spent years feeling like I didn't fit in at my conservative Christian school, and then felt like an outsider all over again in college. But the feeling that I fit nowhere gave me the ability to fit anywhere. Like a chameleon, I could read the newsroom and shift colors and tones, depending on the executive, the reporter, or the scenario I encountered. A child of a messy divorce, I was used to standing back and watching the plot unfold, to understanding the subtext of conversations and what people *really* meant to say. I'd spent my childhood pushing for recognition in a home that didn't leave much room for it, so I knew how to ask the right questions to help someone feel seen and make them feel special.

Deb, on the other hand, was creative and introspective, and interpreted the world through the camera's lens.

One morning, she sat slumped over the news desk, particularly quiet.

"All okay?" I asked, trying to toe the line between prying and concerned.

"I'm fine. I get like this. I have for my whole life," she said, waving a hand.

"Have you ever tried antidepressants?" I asked, going out on a limb. "I used them in college."

She studied me, unsure of what to make of the comment.

Sensing her pain, her dislocation, I invited her over to my place.

That weekend we climbed the five flights to my East Village apartment, and then we went right back down to start the evening at Bua. We ended up barhopping around Lower Manhattan, the next morning waking up with hangovers and stories of strangers we'd befriended the night before. This became our routine, and the weeks passed in a blur. Late nights and early mornings in the newsroom, we bonded, scheming over stories we wanted to shoot, sipping Malbec (her) and gin and tonics (me) after hours. After years of struggling to find where I fit, I felt at home in the newsroom, spending booze-fueled evenings with Deb.

A month into my new gig, on September 29, 2008, I was sitting in the newsroom when it erupted.

"Get Ali on-air!" our bureau chief shouted from his glass office down the hall.

"He's en route!" another producer called.

The Dow was plunging. One hundred points, two hundred points—it just kept dropping. CNN's chief business correspondent, Ali Velshi, an animated bald man with black-rimmed glasses, ran to the camera propped in front of the newsroom desks and started giving live updates.

"Shock and panic on the trading floor!" he exclaimed. His voice grew louder, and he began waving his hands as the Dow continued to drop. Phones were ringing, emails dinging, with alerts including phrases like "historical losses" and "financial crisis."

Suddenly we were in a full-blown recession. Overnight, jobs became scarce, and while I was in one of the most exciting seats in the world, I was stuck in a game of musical chairs; when the music stopped, either I'd get a job offer or I was out. Now more than ever, I felt the pressure to distinguish myself from the other minnows. Guest greeters. Tape loggers. Producer's helpers. Occasionally we might get to write a wire. Six A.M. call times meant we saw each other hungover, and we bonded at strange hours against an intense backdrop only a newsroom could provide. But we all knew that only a few of us would survive. The only way for any of us to go from being a freelance news assistant to being a fully employed one was for someone to leave a full-time position. Uncertainty hung over the news desk, and people held on tightly to their jobs.

The nature of the job was sink or swim, and the tasks weren't always glamorous. But no matter what the request, I smiled and reminded myself how lucky I was to be needed: "Segall, I need you to log two hours of the auto bailout." "Segall, I need you to teleprompt at six A.M. for [insert everyone's least favorite anchor's name here]." But every so often, I got handed a big assignment that made it all worth it.

By March 2009, I was no longer a newbie news assistant. Over drinks, Deb and I had toasted the first wire I'd written for CNN.com, an article about how a small business was trying to make do in the global recession by opening up a topless coffee shop in Maine.

"I know what people want," the shop's owner had explained to me. "People like nudity, and coffee is profitable. Sure, I'd start a coffee shop, but I'd be out of work in a week."

"To boobs and business!" Deb lifted her wineglass, and we clinked, toasting the article, my first time trending on CNN.com.

Later that month, I was in the middle of checking the guest list when my phone pinged. It was the desk manager calling me over.

"Segall," he said with a smile. "Think you can handle the Madoff case?"

"Yes!" I practically screamed. "Whatever you need."

Bernie Madoff had swindled the city's finest out of millions in the largest Ponzi scheme in history. It was a New York–based story under a national spotlight. I might have been low on the totem pole in the newsroom, but I was about to win the gold medal of news assistant assignments.

"Go to the courtroom tomorrow first thing!" the desk manager called out. "Hold a spot for Allan Chernoff. Make sure he gets in and assist with everything breaking."

At 4:00 A.M. the next morning, March 12, 2009, I made my way downtown to Pearl Street, which was already beginning to glisten with the blinking lights of cameras and satellite trucks. In the next hour, the street would continue to liven with media: ABC, CNN, NBC—everyone would be there. I battled dozens of other nobodies like me who were at the courthouse early to secure a spot in the courtroom for someone more important, who'd appear live with news of Madoff's fate as soon as it broke.

That was the easy part; then came the catch. The courtroom didn't allow phones, so as soon as the news broke, Allan Chernoff would have to rush out and relay it to the cameras in an old-school game of Telephone. The only problem: there would be a slight delay from the time Allan left the courtroom to the time he'd be in front of CNN's cameras. After all, he couldn't teleport to our cameras with the news, and he couldn't call the moment it would be announced, because he would be phoneless.

After locking down a spot for Allan, I had a thought: *I won't get a full-time position if I only play by the rules. I have to make myself more valuable than a placeholder.* Once Allan had taken my spot in the courtroom, I found a spillover room with pay phones lining the corridor outside. With just the right amount of maneuvering, I realized that I could hear the speakers in the spillover room while picking up the pay phone outside. *I could use the pay phone to call in the news in real time.* This would mean we could get the news out faster than the other networks. An anchor live on-air could read the news in real time, giving Allan the opportunity to make his way out to the camera and share details about what it was like inside the courtroom. A brilliant plan! But it was a risk. I wasn't sure if it was allowed. What if I was discovered and kicked out? I would probably be teleprompting for Helmet Hair or tape logging for the next few weeks, and I definitely wouldn't secure a full-time position.

My head was spinning as I walked over to the pay phone. *I'm really rolling the dice here,* I thought, as I looked at everyone crowding inside the spillover room.

I waited, trying to focus on my breath and slow my pulse. Then it was time for Madoff to admit his crimes.

"I operated a Ponzi scheme," I heard him say to the jam-packed courtroom. "I thought it would end quickly but it proved impossible."

I dialed the news desk number that connected me to a line full of CNN news producers and writers and began relaying the information.

"I always knew this day would come," he went on.

Then came his plea. He pleaded guilty to all eleven counts. As his words were delivered from the courtroom into the spillover room, I didn't have any more time to think. *Here we go.*

"Madoff pleads guilty to eleven counts," I recounted in real time, my heart pounding as the desk manager on the other end repeated my updates.

My brilliant plan was working until I saw a security guard walking swiftly toward me. *Please don't let me be arrested before I get a full-time job,* I prayed to the news gods.

Mind racing, I decided I'd rather be underestimated than penalized. As he approached, I greeted him like we were best friends, waving with an embarrassingly wide smile. And then I continued with my call.

"Yes, Mom, it's historic! What an unbelievable day!" I exclaimed to the entire CNN news desk on the other end of the line.

The guard paused for a moment, perplexed.

There was silence on the line. A muffled confusion. A planning producer in Atlanta laughed.

The guard put his hand out for the receiver, but he didn't arrest me. I hadn't technically broken a law, and we got the news out fast. By the time I raced outside, Allan was staring straight into the camera, live on CNN, which had already reported the news: "Madoff pleads guilty to all eleven counts."

By the time I returned to my East Village walk-up, the cool winter sun had set behind the fire escape of 126 St. Marks Place, and I was almost too tired to drag my body up the steps. My mind was still buzzing, but my legs had had enough. Flopping onto my twin-size daybed I'd proudly purchased from IKEA, the only one that would fit in my tiny brick corner room overlooking the always-awake street, my mind finally had time to process the day. Breaking news, I realized, was like a giant puzzle. Even if I didn't know the answers, the key was to think quickly and confidently, and, no matter what, to avoid showing anyone that I was afraid.

Of course, I was constantly afraid. I was afraid that I wasn't good enough. I wasn't smart enough. I'd screw it all up. Only Deb knew my secret, because she shared it.

But despite the fear that failure was right around the corner, my survival instincts said keep going—just say yes, and figure out the rest later.

A month later, when the 2009 swine flu epidemic hit New York, a producer walked over to my desk. "Anderson Cooper needs to interview a child or family of a child with swine flu. Tonight. Find one."

"Sure!" I said, and jotted down: *Locate a child with swine flu to go live on* Anderson.

The producer handed me a short list of names of known parents whose children had swine flu. Great! An easy start.

I made my first call and learned that Mrs. X had already been contacted by five assistants from other networks and papers—and she wasn't interested. Apparently, every producer in every newsroom in Manhattan had the same list, and every parent had been dodging their phones, which had been ringing nonstop for days with requests to interview their children.

There had to be another way.

By late afternoon, my notebook was filled with ideas slashed through with giant *X*'s. None of the traditional avenues were panning out. With nothing left to lose, I decided to go to the one place no senior producer would think to go: Facebook.

At the time, Facebook was still a platform for millennials—young professionals, college kids, and high schoolers. Hoping none of the higher-ups would think I was screwing around in the middle of the news day, I began my deep dive. I noticed that a lot of the infected kids went to the same high school, and most of them had gone on the same spring break trip to Mexico. I sifted through their Facebook groups. There were groups devoted to high school basketball, last year's prom, and, sure enough, a group devoted to the spring break trip in Mexico.

Should I join the group? Can't hurt, I thought, adding a number of students as friends, and feeling relief that my lack of inhibition—the spirit that helped me befriend bodega clerks and the hipsters who ran the crepe joint on my block—was translating nicely to my day job.

It wasn't long before I was messaging with a high schooler with the flu.

Hope you're feeling ok, I typed, wondering what an appropriate opening line for a cold Facebook message to a stranger, inquiring whether he had swine flu, should entail. I decided on concerned but not overbearing, and used my youth to my advantage, cracking a couple jokes before convincing the recipient to put his mom on the phone, while I brought in our producer.

Within fifteen minutes, Mom and the producer had a talk, and hours later, Anderson was wearing a protective mask over his perfectly combed hair, in their backyard in Queens, interviewing the young man from afar.

As I watched the interview air on TV from my desk in the newsroom, I had two important thoughts: one, Anderson managed to look good even when covered in a protective tarp; and two, there was something noteworthy about my first real booking having occurred on Facebook.

From there, social media became my tool for investigation. When phones started ringing about reports of a plane going down in the Hudson River, I checked my Twitter feed and saw what would become one of those historical moments in the site's history, a tweet by @jkrums: "There's a plane in the Hudson." Along with it was an image of people walking carefully on the plane's wings to the lifeboat. A producer flagged the tweet to the newsroom, and others raced toward the Hudson to cover the story.

At the time, I was one of the few people in the newsroom with a Twitter account, having nabbed "LaurieSegallCNN" when I first started. I immediately messaged @jkrums, and we followed each other. Within minutes, I'd asked for his phone number and passed it on to our producers. It didn't take long until he was live on-air, speaking about the iconic image he'd snapped and tweeted. It became a touchstone for breaking news.

That tweet represented so much of what I was learning: the power of technology to democratize information, to report, to communicate. Emerging platforms like Facebook and Twitter were beginning to break news, bolstered by the rise of smartphones and the fact that, increasingly, so many of us were carrying them. I had traded my BlackBerry for an iPhone when they debuted in 2007, and after Apple introduced the App Store, I watched as it became a canvas for creative ideas that could reach millions. I was intrigued; but in the newsroom, technology seemed more like a sideshow than the main event.

The hard-nosed journalists I worked for were focused on other things: the recession, President Barack Obama's first year in office, the sudden death of Michael Jackson, and the balloon boy hoax.

I learned to negotiate interviews from Dara, a producer with a reputation for being kind to the news assistants.

"We'd love for you to come on and share your story," she'd say to a mourning mother. "But I don't want to force you to do something you're not ready to do." There was empathy in her tone. I knew she meant what she said.

Other producers taught me to triple-check my sources.

"How do you know he's credible?" Alex, a producer who'd been a staple of the news desk for years, pushed me when I brought him a tip from a source. "We need more to go off of. Keep going."

I learned to work quickly, but cleanly. I learned how to pick up the phone instead of relying on email; how to check the facts until I was certain without a doubt; how to pursue a story relentlessly and with integrity.

But my year at the Desk was coming to an end, and unless a position opened up, I'd be jobless, stuck with rent I couldn't afford, my dreams of becoming a journalist crushed.

Then, in what appeared to be a bit of luck, the news desk announced a rare opening: they needed a Tapes Desk coordinator. The job was to take in all the incoming feeds, give them media source numbers to ingest into CNN's system, and feed them out. I'd worked as hard as anyone and thought I had a good shot.

"I bring a creative approach to the news, and a young approach to a constantly changing industry," I explained, listing my accomplishments to the executive grilling me.

It had been nearly a year since our phone interview that landed me at the Desk. Now Gene, the managing editor of the newsroom who oversaw new hires, nodded enthusiastically as I passionately stated my case. I felt good. My instincts told me the interview had gone well.

That night, Deb and I went to celebrate at Bua. Fionne, a charismatic Irish bartender and neighborhood heartthrob, was returning to Ireland the next day, his visa no longer valid. One of the locals, an old guy who proclaimed his ardent love of *Star Trek*, told the group he was moving to Bushwick. A dive bar across the street had shut down, to be replaced by another. Our local group was beginning to disperse. Even though I had only lived on the block for less than a year, I could already feel a transition happening as a new wave of incoming hopefuls came through, marking the always-changing tide of the place I had begun to know as home.

Fionne, who was on his sixth whiskey, was growing increasingly sentimental with every shot. He looked over at me, draping himself across the bar, slurring his words. "You're going to be anchoring CNN by the time I get back," he said with drunken confidence. I looked outside, watching as Maria cruised by in his red salsa dress, preparing to make his entrance.

I looked at Fionne and laughed. "Yeah, right," I said, taking a shot, numbing myself to the idea of logging hours of tape the next day. "They wouldn't put me on-air in a million years."

I looked outside the cherrywood bar. The Village was alive with budding trees and guys in skinny jeans and floppy haircuts. For a moment, I was optimistic.

"It was a tough call," Gene explained three weeks later, as I saw my dreams of staying in the newsroom shatter.

"I understand," I said, without a shred of comprehension. I wondered if I'd have to go back to Atlanta, give up everything I'd been working toward. I wondered if whoever lived in my apartment next would enjoy Maria's late-night serenades as much as I did, or his increasingly destructive drinking habits that led him to climb from his fire escape to mine and pound on the window around 3:00 A.M., yelling, "Sweeeetie!"

I wasn't ready to leave the city or the community I'd started building. I thought about my evenings with Deb, our late-night adventures. We'd just convinced nearby Cafe Centosette to display her photos on the exposed-brick walls of their failing pizza joint on Second Avenue. I couldn't leave St. Marks. I couldn't leave CNN. I felt a sense of belonging and couldn't imagine it being pulled out from under me.

Rejection without perspective felt like the end of the world. Even though the job would have kept me glued to a monitor looking at video feeds of congressional testimonies, I felt as though I'd just been told that I was unlovable, inadequate, and incapable of keeping a front-row seat to history.

Depressed and desperate, I exited Time Warner Center, taking in Columbus Circle, wondering if this would be my last view of the office I'd come to love. I could hear the clicking of the horses' hooves, carrying tourists in red carriages through Central Park, which now would never become my regular lunch break spot. I wandered down the sticky New York streets, passing bodegas filled with three-day-old flowers and the neon signs in Times Square advertising Broadway shows. I hung a right, away from the crowds, letting myself roam aimlessly. Eventually I stopped in a Brookstone store, where, between the massage chairs and gadgets, I saw a display of African dwarf frogs floating in stacked plastic containers, an odd find for a store known for electronics.

A pet. A living, breathing thing I could take care of. I thought that if I could keep a frog alive, it would prove that I was a responsible adult.

On impulse, I bought two. I named one Joan (Didion) and the

other Travis (because the name wasn't tarnished by any of the guys I knew). Armed with my two frogs and the promise of adulting, I no longer felt like the Tapes Desk rejection was the end of the world.

With just a few weeks left before time ran out and my yearlong freelance position would come to an end, and with no knowledge of the financial markets, but determined to get back into the newsroom, I applied for a job that had opened up in Business Updates, which was responsible for providing market-move segments for the day. This felt like my last shot. There was an out-of-sight, out-of-mind mentality once you'd left the building, and if I walked out of CNN without a job lined up, chances were I wouldn't get back through those revolving doors.

After filling out an online application, I was asked to take a business news quiz. Hunched over a desk a floor above the bullpen, I filled out a test devoted to market moves and knowledge of the stock market. As I guessed at questions I had no idea how to answer, I mentally scribbled my short-lived media industry obituary: *She came, she teleprompted, she failed.*

I was shocked when a week later I received miraculous news: I was accepted! I couldn't believe it.

I raced downstairs to find Deb sitting at a desk, looking bored.

"You won't believe it!" I said, out of breath.

"Tell me you got it."

"I got it!" I said, trying not to raise my voice on my old turf.

"How did you know so much about the stock market?" she wondered.

"Either luck on the quiz, or they just really needed to fill the position," I said. Either way, a year after freelancing and dancing around commitment I so desperately wanted, I could breathe. I had become "official" with my dream job.

A week later I was the new production assistant at Business Updates. Unbeknownst to me, I'd landed at what was deemed the "bad wedding table" of CNN. It was led by a guy named Stan, who couldn't get anyone's name right. Stan called me Lauren or Mackenzie, depending on the weather.

My main task was to write explainers for anchors to read on the daily market moves. I needed to find people who could help me understand the world of finance, so I browsed CNN's internal guest-booking catalogs for cell phone numbers and contact information of the most powerful hedge fund managers and started with the basics: "I'm a reporter with CNN. Obviously, I get what bonds are [this was not entirely true], but how would you describe them to someone who doesn't quite understand?" Those three letters—*CNN*—held weight. And as long as I didn't make myself out to be a total rookie, or proclaim myself a production assistant, they wouldn't know how low on the totem pole I actually was. I'd piece together their explanations and type them up for anchors to read in their Business Update segments.

I began learning how to write television segments and the banners that appeared on TV below anchors as they spoke, and how to code them into the system so they appeared on-air. The hours were much easier than when I was at the Desk, but I found writing about the markets for anchors mundane, and I rarely escaped the building to assist on shoots. It didn't take long to do the writing tasks, so after I submitted my scripts and pulled whatever tapes were needed, I joined the other production assistants to watch *Real Housewives* on our TV monitors. I finished at 4:00 P.M., with the closing bell, and then I went home to my apartment.

With my evenings and late afternoons now free, I had the time to pursue my budding interest in technology. I stumbled on a tech meetup through a mutual friend and from there started going every week and getting to know people who were creating apps, which were beginning to take off. Even though no one knew where they were going, I felt pretty sure they were going somewhere. I'd grab a drink after work with some oddball who was building a new company, and at eight the following morning, I'd return to Business Updates with new information that I didn't know what to do with.

The more I wrote about Wall Street, the more I dreamed of the job I really wanted—writing what I viewed as the next iteration of the corner stories I'd covered in college: lesser-known stories of people innovating on the edges, the ones far from the glare of Wall Street. I wanted to cover the entrepreneurs I was meeting in dive bars and tech meetups. I wanted to write about the people no one else was paying attention to, the ones the world had yet to notice.

Wall Street wasn't exactly cool anymore. The shine was wearing off; and as the market and the Madoffs took a dive, the American Dream took a hit, too. But there was a silver lining. In 2009, out of the ashes of the recession, a new creative energy was emerging from the scarcity of that moment. A space for the misfits was created. They were smart and ambitious, but didn't conform to the established norms set by corporate America. They weren't ruled by the "shoulds," and weren't convinced the world had to be a certain way. They were entrepreneurs who didn't want to fit into the system: they wanted to build their own structure. They were outsiders who saw the world differently, and their world felt like a place where I finally belonged.

CHAPTER 4

Fake It Till You Make It

Still groggy from late nights in jam-packed bars, drinking too many gin and tonics and talking about location apps, I pulled out my phone and typed: *How'd the date go? Was she Boardwalk?* Then I pressed *send,* and the message flew through the pipes.

Definitely not Boardwalk, Daniel typed back.

Daniel was my New York family. When I first arrived in the city, I'd been a starry-eyed twenty-one-year-old intern who believed Times Square was almost as magical as the view of twinkling Manhattan skyscrapers from my apartment's fire escape. I'd immediately bought a street painting in SoHo, played chess in Washington Square Park with a man named Louise, and discovered a wonderful thing called birthday cake ice cream. I gave more money than I had to spare to my homeless neighbors.

I'd walk down MacDougal Street, laptop bag draped over my shoulder, and head straight to my favorite writing spot, Esperanto Café—an Israeli coffee shop, open 24/7. The service was terrible, but the conversations were unforgettable. People roamed in and out, sitting at tables too small, seats too cramped. A velvet couch, framed by neon lights, overlooked the street, where promoters dropped flyers for comedy, music, and art, and twentysomethings wobbled in and out of Panchito's Mexican restaurant to a street that smelled of curry, hookah, and trash. One day I would sit next to a man reading Russian literature, and the next, I'd run through a script with the actress beside me. We all came from different corners of the world, but something about the place allowed us to talk to each other in a way we wouldn't elsewhere. Maybe it was the vicinity of the tables, the lack of privacy, or maybe the coffee shop just practiced what its signage preached—Esperanto, a universal language. We all spoke it. We all felt it. Whatever it was, I drank it up. Surrounded by tiny wooden tables and worn couches, I hoped to find a shared idea, or purpose, or dream.

I was still looking for my place, eager to connect. I still made eye contact with people I passed on the street, something veteran New Yorkers would chide against. But I believed that somewhere in the humming and honking, I would eventually find belonging. At the time, I had no idea Esperanto would close down and be replaced by a series of chain stores, and soon the only neon lights I'd gaze at would be in the palm of my hand. I had no idea that a new universal language was about to disrupt the entire world, and that I'd be at the center of it all.

One evening, sitting at a dimly lit bar called Freemans on the Lower East Side, I started talking with a woman named Talia. She was New York cool, with long blond hair and kohl-lined eyes. I put her number in my phone, convinced she would be my new best friend. A couple days later, I sent a message to the number: *Let's meet up! Can't wait to see you!*

We scheduled a drink, and as I waited at the bar, I was approached by a guy I vaguely recognized as the cousin of one of my best friends growing up in Georgia. He sat down and stuck out his hand.

"Glad you reached out," he said, his shaggy brown hair dancing around his big, boyish eyes. "I'm Daniel."

I stared at him blankly and said, "I thought you were Talia."

"What are you talking about?"

Somehow, I'd mixed up the numbers, but I soon forgot about Talia, and Daniel and I became fast friends, meeting nearly every week at Cafe Centosette. Usually we were the only patrons, but we considered ourselves VIPs at the always-empty bar.

I came to realize that Daniel was the kind of guy who only cared about a few people, but he cared about them with Mafia-like loyalty. A charismatic chain-smoker, he was successful in his own business. He dated a lot, but never really opened up. Few people knew that he'd watched his dad, who was an alcoholic, die of a heart attack.

Despite that, or because of it, he took on life with an extraordinary amount of energy. He was like a wave that hit you before you knew it was coming. Daniel's smile lit up a room, drawing people to his larger-than-life persona. He navigated the city like it was a video game, always leveling up, jumping over obstacles, beaming all the way to the top while fighting the bad guys.

Even so, we shared the same insecurities. Like me, he had the same deep-rooted fear that he wouldn't "find normal," and that somehow, he wasn't deserving of it. So he hustled, graduating early from NYU and building a career in real estate. We always made sure we were in on the joke, because we feared the joke was on us. We made light of the day's absurdities, toasted our clumsy attempts at dating, finding a relationship, and creating a life that was *spectacular.* We called it "Boardwalk"—the concept I'd adopted growing up in the leafy suburbs of Georgia.

The meaning had evolved since my middle school years into a search for something greater: the idea that something spectacular was just around the corner if you refused to take the easy route or give up. Boardwalk was a relationship that didn't end up in destruction, one where both parties refused to settle. Boardwalk was the search for a community that didn't feel stifling or judgmental. Boardwalk was spectacular. Though I was no longer fifteen pounds overweight or dressed in oversize flannel sweatshirts, I was still on my quest for something better—something extraordinary. I sensed that same feeling as I met tech entrepreneurs who refused to accept a mediocre world. They wanted a better one—something beyond what was in front of us.

By 2010, a tech movement was officially brewing. Thanks to the success of the Apple App Store, a new class was emerging: scrappy, optimistic, out-of-the box entrepreneurs who were far from the debris of Wall Street. I lived for my late-night meetings with the people entrenched in a growing tech community, and I started to wonder if I could turn my very-off-the-record conversations into something more than a curiosity. When my brother, a creative guy with an eye for the cutting edge, told me technology was the future and I should be paying attention, I knew I was onto something. Back at Cafe Centosette, I practiced pitching my ideas to Daniel and Deb and schemed ways to produce tech segments on the side.

"What if I just tried interviewing tech entrepreneurs in my free time?"

"You're done at four P.M. every day," Daniel chimed in.

"I'll shoot, you produce," Deb said, sipping her Malbec. "We'll make it easy for them to say yes. They won't have to do anything."

"Stan will never let me do it," I complained about my boss at Business Updates. "He still calls me Mackenzie."

"Why not ask Caleb? See if he'll let you produce tech segments as a side gig."

Caleb was the head of CNNMoney. He wielded more power than Stan, and if I played the corporate chess game correctly, perhaps there was a chance I could take the first step toward making my dreams come true and leveling up in the newsroom. I may have been a mere production assistant with less than two years of training, but it was becoming clear that the job I had was just a placeholder for the position I *really* wanted.

The next day, battling a hangover and a large dose of self-doubt, I made my way toward Caleb's office to make the sell.

Let me interview one entrepreneur, and if you hate the segment we produce, I'll never ask again, I silently repeated to myself, walking down the aisle of the chirping newsroom to Caleb's door.

A young manager with a good sense of humor and cool-rabbi vibes, Caleb listened as I recited my talking points: *I'll bring you interesting tech interviews. Deb will shoot them, so we won't need resources. You don't have to do a thing. Just say yes.* "We'll start with one. You hate it, and we can pretend this never happened," I said, delivering my final line, unsure where to put my hands.

He hesitated, looking at me like I was a talking squirrel. I had straightened my hair and worn my most professional attire: a black fitted blazer and skirt. I'd hoped it read "power suit," but as I waited for his response, it seemed to suggest "attending my own funeral," and I began to lose hope.

But then he cracked a smile and said, "Okay, Segall. Let's give it a try."

"He's letting us use a crew car?" Deb asked the next day, her eyes wide. In a stroke of luck, or clear lapse in judgment, Caleb had agreed to let Deb and me borrow one of CNN's crew cars. Normally they were reserved for professional photojournalists and people who knew how to drive them.

"Honestly, I can't believe it either. But there's no way I can drive through New York."

The last time I was behind the wheel was over five years ago, in the suburbs of Sandy Springs, Georgia. My license was well on its way to expiration, and I could think of nothing more alarming than weaving through New York traffic with all the crazies racing through yellow lights, honking at cyclists, and yelling out their windows.

"I can try," Deb said, unconvincingly.

I'd booked an interview with Biz Stone, one of the three cofounders of Twitter. Even though the mainstream media had yet to catch on, the company was on its way to becoming one of the hottest social networks in Silicon Valley.

"He thinks I'm a producer," I explained to Deb. "Do we think that's a problem?"

Biz and I had exchanged a couple of emails, in which I'd introduced myself as "Laurie from CNN." I didn't *lie*, but I relied on ambiguity to avoid revealing my lowly status as a production assistant.

"What's the phrase?" Deb smiled and grabbed the car keys. "Fake it till you make it!"

With the radio blasting, windows down, and my Converse sneakers propped up on the dashboard of the passenger seat, Deb drove the crew car to the East Village for the interview with Biz. We chose a red bench on St. Marks and Avenue A. By then, I'd moved two blocks over to Eleventh Street, into my first studio apartment, but I loved the idea of honoring the occasion by shooting the interview on my old block. It was good for visuals and had a downtown vibe. I couldn't think of anything worse than putting a tech geek under the bright lights at CNN's studio. Biz, and most of the tech people I spoke to, didn't fit in against the glaring light where celebrities, politicians, and anchors sat. They didn't speak in sound bites—their sentences were often too long, unrehearsed, and brainy. The studio felt stifling for the free-spirited nature of the tech movement, and our interview setting needed to reflect that spirit.

I met Biz at a local Italian restaurant, so I could have a short talk with him before the interview. Essentially, that meant a couple of self-deprecating remarks on my part to get him to open up and to give Deb time to set up the shot.

"Hey!" Biz greeted me with ease. He was young, seemingly open, and unaccompanied by any PR reps. If anything, he felt like the kind of stranger on a plane who'd strike up a conversation.

Ten minutes later, I got the text from Deb: *Let's do this.*

"Go time!" I said, turning to Biz as we walked a block to the red bench.

We sat in front of Deb's camera, and Biz cracked a joke about smoking pot and how his hippie wife let wild animals roam the house. I smiled in what I hoped was a professional way, while picturing skunks and turtles wandering around a marijuana-filled home somewhere out West. From the bench, Biz and I waved as a tour bus drove by. Its passengers looked at us, likely wondering who this man in a black shirt, jeans, and wire-framed glasses was. No one knew he'd eventually be worth hundreds of millions of dollars.

"I'm rolling!" Deb called, pressing *record,* no doubt rooting for me to appear more relaxed than I felt.

"So where'd the name Twitter come from?" I asked Biz, feeling Deb's camera on me, aware that I was lucky to be on this side of it. It was unheard-of for a production assistant at CNN to appear on camera. Normally it took years of reporting in local news markets before getting in the door at a national place like CNN and securing a spot on-air. I had never worked in local news, and until recently, hadn't thought about appearing on-air. Biz didn't know any of this.

I laughed and spoke with what I hoped was assurance, but I was painfully aware of my own voice, how I moved my hands, and where I should place them. I knew I'd managed to game the system, but I couldn't seem to fake the years of experience.

"We had a short naming session," Biz replied. "There were words like 'Jitter.'" He laughed, adjusting his glasses.

"So we would be *jeeting*?" I said, going for some light humor before pivoting into more substantive material. "You're an idea guy," I continued. "How do you gauge potential in an idea?"

"The easiest thing—is it resonating personally with you?" he said. "Is it something you really want to work on, despite people telling you it's stupid and doesn't mean anything?"

I'd heard the sentiment before. Many of the entrepreneurs I was meeting had the same story. They'd been hit with skepticism from the industries they were trying to disrupt; veteran employees, people who were stuck in their ways, didn't understand what they were trying to do. But these entrepreneurs ignored the external noise telling them their ideas weren't any good; that they wouldn't succeed; that the world was *supposed* to look a certain way, and we were *supposed* to do things in a certain way because that was how they'd always been done. Of course, most people didn't like change; change wasn't familiar.

But I'd always felt like a stranger to the familiar. I knew what it was like to have parents waging a legal battle instead of sitting together at the dinner table; to feel like an adult before I'd managed to master puberty; to spend high school apart from my only brother, who was at boarding school the majority of my high school years.

I wasn't sure what normal was, but I knew I hated "familiar." "Familiar" was the women who ignored my mother at the bus stop because divorced women were branded with a scarlet letter. "Familiar" was the beer-soaked conversations with Kappa Alpha Theta on Washtenaw Avenue, the cruel rumors that echoed through the royal-red halls. "Familiar" was painful and boring. Technology and the people creating it were the opposite of "familiar," and I couldn't have been more on board.

As we wrapped the interview, Biz looked down at the bench we were sitting on. Scribbled onto the red-painted wood were names and love proclamations.

"Should I leave my name?" he joked, pulling out a pen and scrawling his name on the bench, leaving an analogue of himself behind in the digital world he was guiding us into.

"This isn't bad!" Caleb said, leaning back in his chair after watching the piece Deb had cut together, titling it "The Biz of Twitter." It was released on CNNMoney.com that day.

That night Deb and I toasted our success at Cafe Centosette, and then kept scheming. The next stop would be SXSW.

SXSW (South by Southwest) was a conference that had been around for ages, but whereas in the past it had been all about music, it had recently started leaning into emerging technology. In 2007, Twitter got its first publicity bump at South By; in 2008, an interview with Mark Zuckerberg devolved into chaos; and in 2009, Dennis Crowley launched a hot new startup called Foursquare. By 2010, anyone in the tech world worth paying attention to went to the conference. If a startup "won" South By, that was it. And if it didn't, well . . . it was sink or swim.

Not long after, I sat next to the control room at CNN, dialing Foursquare's office phone. Once I'd proven myself with a couple of test runs, including the segment with Biz, Caleb had started allowing me to produce a handful of segments for CNNMoney. But my *actual* job was still production assistant at Business Updates, where I was writing daily updates for news anchors. I'd started slipping tech pieces into the market updates, writing segments for anchors that included some of the startups I was encountering in my off-hours. I'd just written one of those segments on Foursquare's partnership with Bravo, and found the company's office number in a long email chain. I called the number to research another story I was working on: burgeoning tech companies heading to Austin for SXSW.

"Hello?" a muffled male voice answered.

"Hi! Laurie from CNN."

"Oh hey. It's Dennis."

"Crowley?"

I was shocked that a founder answered his own company's phones; I'd imagined layers of PR. After a couple minutes of small talk, I got to it.

"What's going to be hot at SXSW this year?"

"It's going to be *us*," he replied matter-of-factly.

"Are you just saying that?" I responded, twisting the phone cord around my finger. *Is he cocky or confident? Both?*

"Wait and see. Are you coming to Austin?"

"Definitely!" I said, although "I hadn't thought about it until this very second" was the more honest answer.

Later that evening at Centosette's empty bar, Deb and I stewed over the dilemma.

"CNN wouldn't send you to Austin if your life depended on it," she said.

"What if I just said I was already going to be there?"

"Can you do that?" she asked.

"I may not be able to afford my rent this month, but otherwise, I think it's worth a shot."

We devised a plan. I'd take a couple days off from Business Updates and call it a "vacation." Then, I'd tell Caleb I was going to be in Austin for the conference anyway. I'd try to convince him to pay for Deb to come along and shoot, while we produced stories on the ground. We would share a bed in a cramped motel room and shoot four, five, six interviews a day. I just needed to set it up as a win-win scenario for him.

Incredibly, the pitch worked: Caleb agreed to send Deb to accompany me.

As soon as we arrived at our hotel, it was clear that Dennis was right. SXSW 2010 was all about the battle of the location apps.

Led by Dennis, Foursquare, along with its competitors Gowalla and Loopt, were gaining traction as people used their smartphones to see where their friends were and to earn rewards. As Deb and I roamed downtown Austin's music-and-barbecue-filled streets, we watched techies battle it out in "check-in wars," virtually announcing themselves at coffee shops, bars, and parties that lasted into the following morning. In a world where the hot new thing was created by twentysomethings, the spirit of SXSW was competitive and playful, and beer flowed day and night.

Pretending to be a producer, I lined up interviews with a group of ambitious, relatively unknown young founders. While many of their companies would soon turn into successful endeavors, at the time it was hard for them to get on the radar in traditional media outlets, which didn't yet see the power and influence of startups. So, while I may have been a young wannabe producer, I was also one of few in the mainstream media paying attention to them then, and they were happy to speak openly about their ideas.

When Deb and I arrived at our interview with Dennis, he wore a Foursquare-branded T-shirt and looked hungover.

"Is he actually playing Four Square?" Deb said to me, gripping the tripod as we approached the startup king.

I glanced over to see him bouncing a ball on sidewalk chalk. Someone had drawn an actual Four Square board on the concrete.

"Affirmative." I looked down at my chipped nail polish, and realized we'd be shooting B-roll—extra footage we'd use to cover the interview—of me scrolling on the app. If I was going to play producer, I chided myself, I had to stop biting my nails. My apartment might look like a crime scene, and I might have a knack for losing important documentation like IDs and credit cards, but I needed to hide my tells and appear as though I had it together.

"SXSW is like spring break for nerds," Dennis explained twenty minutes later, on a bench outside the convention center. As Deb tilted the camera at him, he looked toward me and described how his location technology was leading us to have "social awareness of what everyone's doing." With "check-in" apps, we could see where our friends were and what they were up to, and create a more social world.

Next up was the cofounder of Loopt, Sam Altman, an exuberant whiz kid from St. Louis, with light brown hair, light quizzical eyes, and a brain that seemed to always run on high speed. Unlike Dennis, Sam was quiet and introverted.

I stood in front of him, asking questions, but because he was sitting, we couldn't seem to get him to look into the camera.

"Just look toward Laurie's chest when you answer the questions. You'll be eye level with the camera," Deb instructed.

Sam's face went from peach to beet red.

Dear Lord, could we at least act professional?

As Sam followed the instructions, I wondered if all of us were just playing grown-up.

The days bled into nights, and Deb and I mixed hustle with hangovers, sneaking into parties, attending dinners with venture capitalists (VCs), dancing under the stars, and meeting rising founders. I interviewed Josh Williams, Gowalla's founder, and then Twitter's CEO, Ev Williams, a soft-spoken, philosophical guy.

"You're reading an article on the Huffington Post and the Flaming Lips are mentioned, and you can, from right there, follow the Flaming Lips on Twitter," he explained, puffing up about Twitter's growing influence.

Later that evening, someone leaned over and tipped me off that Ev was being pushed out of the CEO position at Twitter. I thought back to his quiet enthusiasm and wondered what he hadn't said in our makeshift interview. Not having enough to go on, I didn't publish the information.

The last night of the conference, we took a break from editing and interviews to go to Foursquare's party at a live-music venue called Cedar Street Courtyard.

"You're sure you can get us in?" Deb asked nervously, glancing at a large security guard who determined which guests gained entrance to the coveted VIP section. I was pretty sure our names were nowhere on the list.

"Nope, but the worst thing we can do is show fear." I smiled as she tucked the camera into her bag. We approached the security guard with a false aura of confidence that was becoming increasingly comfortable.

"Hi! How are you?" I said enthusiastically, making eye contact with the large man guarding the entrance. *Make a personal connection.*

He nodded, acknowledging my presence. But the moment of truth was about to come.

"Name?" he said.

"Laurie Segall . . . from CNN," I said, self-assured, adding, "They said I had a plus one."

I could feel Deb shifting uncomfortably behind me. I positioned my body in front of hers, blocking his view of her. She had a terrible poker face.

He sifted through the papers.

"I don't see your name here," he said.

"Oh wow, that's so weird. I'm so sorry. I can contact them," I responded coolly, having absolutely no idea who "them" was. "They'd told me I could bring someone. I'm so sorry for the misunderstanding," I added again, waving to no one in particular inside.

I saw the security guard glance up. He was contemplating his next move.

Please let this work.

"You know what? Just go in. Don't worry about it," he said, handing me and Deb two VIP bracelets.

"Thank you!" I replied, walking in, hesitating for a moment to think about the absurdity behind all of it. The line. The list. The chosen ones who got to call themselves VIPs. I heard the man behind me identify himself as a "Twitter celebrity" to the security guard and I exhaled, rolling my eyes and walking up toward the crowded balcony.

"That was stressful," Deb whispered, heading straight for the bar.

Inside, Foursquare logos were projected against climbing ivy on concrete walls, and twinkling string lights lit a space that felt like an adult tree house. Everyone from entrepreneurs to self-proclaimed Twitter celebrities in bright hoodies and lanyards bobbed their heads to a DJ set. Ashton Kutcher stood around nonchalantly while nerds pretended he was one of them. People later floated over to a suite at the W Hotel for Kutcher's invite-only jam sessions, where entrepreneurs stayed up all night drinking whiskey and discussing the future.

On the final morning of the conference, I woke up with a hangover and inspiration to keep moving forward. I had spent nearly all my savings, but it was worth every penny. The videos Deb and I had created during the conference that landed on CNNMoney.com were performing well. Data showed people were clicking on our videos, and the ones we'd piece together once we returned home would also do well. The question of whether people were interested in startups and the emerging tech beat was answered, and I had a feeling this trip would give our bosses the evidence they needed to back up my thesis: this was the future.

When we arrived back in NYC, Caleb agreed to send us to SXSW next year, on CNN's dime. As I was turning to leave his office, he said, "You know, Segall, I'd roll the dice on you."

It seemed to me that I'd figured out a winning formula: Convince someone you know something. Do the legwork. Get a couple of somebodies to start mentioning your name, and boom—*you're* somebody.

By that point, there was a growing group of somebodies. Along with a couple other young producers, I'd successfully convinced someone somewhere I was important enough, and like wildfire, names were dropped, invites rolled in, and we were starting to be viewed as the young people who "got it." We had an idea of what the future looked like.

That summer, New York was boiling hot, but tech was cool. Dennis appeared in subway ads for Foursquare, and he was plastered across the covers of *Wired, Forbes,* and *Fortune*. The New York tech scene was booming as advertisers, entrepreneurs, and creative types schemed about projects and partnerships. Daniel and I filled the evenings with gossip, talk of failed dates, and our latest work ambitions. I'd wake up after only a couple hours of sleep and roll out of bed at 6:00 A.M. As part of a seasonal art exhibit, pianos were scattered throughout the city, and before making my way to the newsroom, I'd play songs from *Les Misérables,* the notes echoing on the empty East Village streets, before the vendors were up to sell fresh fruit.

As I typed up scripts on the stock market, I thought about the tech founders who had relationship issues and stayed out too late. The ones who seemed to have a soul. I thought about the ego-driven ones—and ego was rampant, because everyone was in on the secret: We were part of something big. Something that was changing the world.

CHAPTER 5

Finding My Superpower

I was standing in the Time Warner Center next to the escalators that go down to Whole Foods, thinking the lines were too long and the food too expensive for my Business Updates salary. I had just left work and was on the phone with my dad. Ever since our purity pledge ball, we'd started speaking more frequently, having phone calls every couple of days.

My relationship with my father still had all the tentacles of the pain and dysfunction of my parents' divorce, but the years had softened the blow and eased some of the anger I'd held within me.

"I've got to get a second job, Dad," I said, rattling off options. Waitress. Bartender. Anything to supplement living out my New York Dream.

"Laurie, I had a headache."

I rolled my eyes. I was talking about my *future* and he *had a headache?*

"They think it's brain cancer," he said simply. My dad the doctor revealed his own diagnosis: a tumor on the left side of his brain.

His words were like bullets, their impact causing me to crumble against a nearby statue of a large naked man sculpted by a well-known artist. To this day, it mainly attracts tourists, laughing and taking selfies while fondling key anatomical areas. But that day, the enormous twelve-foot statue was a metaphor for the massive news that would change everything in my family.

I leaned against a giant bronze leg as I struggled for air and listened. He said that his brain was bleeding. He would get surgery in two days. They were lucky they caught it.

When we hung up, I called CNN, dialing Ross's extension. Within seconds, he was on the other end of the line.

"Ross, I can't breathe," I managed to say in between sobs. "It's my dad. He might die."

There was a pause.

"Where are you? I'm coming."

Ross and I walked side by side in silence down Ninth Avenue, ten, twenty, fifty blocks. I thought about how many years I'd spent angry at my dad for divorcing my mother, for not calling often after they'd parted. I thought about the phone calls he *did* make, the ones I'd avoided because I was angry; now missed opportunities, precious wasted time. It hit me: I had taken on much of my mother's pain. But the moment I hung up the phone, my anger was dislodged. It was replaced with immense pain and regret for the years we might not get to have together.

As Ross and I neared Chelsea, I wondered if my father would ever walk me down the aisle—a thought that surprised me, given that I'd never even let myself envision finding someone, much less marrying. For the first time, I allowed myself to want him by my side, to feel what so many daughters feel: I just wanted my dad.

I let the fear wash over me with the sounds of the busy city street. As an ambulance drove by, its red lights bouncing off the buildings, I flinched.

"What if he's in pain? What if I lose him? What if he dies?" I spoke with quiet hysteria.

Ross didn't tell me everything would be okay. He hugged me, and we parted ways.

I immediately booked a flight home. Before I left, I emailed Stan a note that I was shocked to have written: I told him I needed to be there for my family, that my father might not live, and that depending on how my trip to Atlanta went, there was a chance I wouldn't return anytime soon.

I hit *send*, surprised by the part of me so willing to give up my path to take another one—focused on healing and spending the last moments with someone from whom, up until this day and this call, I'd felt disconnected.

How quickly it became clear that as the world collapsed around our family, I wanted to spend every moment with my father. I wanted to make up for lost time.

As I threw clothes in a bag, I dialed my aunt Sharon's number. Sharon was like another mother to me. I'd worshipped her growing up. My father's sister, she was beautiful and always impeccably dressed, her shoulder-length blond hair framing her slender face. But most of all, she was safe.

She'd offered me comfort during moments in my childhood when I wasn't sure whom to turn to, when I was reaching for a parental figure. She'd sat with me in the bathroom during my father's wedding, when I'd burst into tears. She'd rubbed my back, wiped my tears, and allowed me to be a child with feelings. She didn't leave my side that day. I wasn't alone.

"Laurie," she answered immediately. "How are you?"

I couldn't get a word out, just sobbed.

"I know," she whispered.

I envisioned her clear blue eyes filling with silent tears. She was always stoic, even in her compassion.

"Laurie," she said.

"Yes?"

"He needs you."

I knew what she meant. My father was facing the most terrifying moment of his life. I could give myself permission to love him the way I needed to be loved, and that wouldn't come at a cost to my mother. I had always felt a fierce loyalty to my mother, as I'd watched from the sidelines the unraveling of her reality: the legal battle with my dad, the bursting of her bubble, and the domino effect of chaos that ensued that had taken parts of her away, too. And while she would always be my emergency contact, and the first to overreact if I felt even slightly unwell—owning her "Jewish mother" status—I always treaded lightly around her with news of my father. There were simply too many land mines.

Late afternoon of the next day, I landed in Atlanta. The sun sat low in the sky, forcing me to look as it got smaller and smaller. Why was it that in the face of mortality, there was beauty everywhere? The sunset was never as clear, the fall leaves never so bright.

That evening I arrived home at my mother's, heartbroken and conflicted. My father and mother had rarely spoken since I was in college, but my mother was respectful and sat with me at the kitchen table while I cried. "They still don't know the diagnosis," she said, reaching for words. I could tell she felt a different type of pain hearing about my dad's condition. But it was hard to exchange our collective experiences. The turf was still radioactive.

I walked into his hospital room at 9:20 the next morning. I held my breath, terrified, not knowing what to expect: How would he look? Would he be able to speak? To hug? I was comforted that at least he'd be getting the best care. After all, he was a beloved doctor here at this very hospital, where they'd opened up his brain.

Dad was sitting up, eating French toast with too much syrup, and watching CNN—a Business Updates segment I'd normally help produce. Half of his hair was gone, and a big bandage covered a shunt that would alleviate pressure on his brain. It would help the headaches and the double vision.

He was smiling, the top half of his robe undone, looking strikingly vulnerable, but still asserting authority. He looked smaller than I remembered. I let out a sigh of relief, watching his eyes light up as he told the nurse the benefits of getting the swine flu vaccine. Despite the circumstances that felt so unfamiliar, he still seemed like himself.

Don't cry, I whispered to myself, hearing my aunt's words from our conversation the night before: "He needs you."

His new wife, my stepmother, Harriet, sat in a chair in the corner. She'd been sending out updates to the family since the Call, and I was grateful. My brother was flying in from Argentina, where he was living. When the News had come, there'd been no question that we would return home. My brother and I both wanted desperately to be there, to heal and move forward together.

For now, it was just me and Harriet and my father, who appeared to be in good spirits. Today was the first day I'd allowed myself to see Harriet as my father did: as a supportive wife and partner. Today she stopped representing my dad's absence, and instead was someone with whom I could build a relationship. We were all relying on her to take care of my father.

I shut my eyes and envisioned a world where I hadn't trained myself to no longer need him, where I turned to him, instead of Ross, with job updates and boyfriend drama. I was filled with anger and resentment—anger for wasting time, resentment for the pain that had guided me away from him. Regret crept through my body, pumping through my veins.

Within the week, Dad came home as we all awaited more details: How big was the tumor? What kind of tumor was it? He had yet to be diagnosed. In the meantime, my brother arrived, and we spent long hours at my father's home, the one that once felt so far away. It almost seemed like the leaves changed colors before our eyes, shifting from forest green to radiant red, while we waited for any news. We watched *The Big Lebowski*; he did tai chi; we ate chicken soup. I felt more whole, more okay, than I'd felt in years.

"I'll be back soon," I said, holding back tears as my father and I walked slowly side by side down the driveway. I couldn't stop looking at the surgically placed shunt in his head, the bulge that showed me his brain had bled. I reached out to hug him, wrapping my arms around him, an unfamiliar feeling. He squeezed back. I was filled with gratitude—painful, angry, joyous gratitude for life. We stood in silence until I let the words come out through my tears: "I love you so much, Dad."

I returned to New York, to CNN, to the cameras and microphones and stale cups of coffee with the understanding that at a moment's notice, I might pack up my bags and return to spend the rest of my father's days with him.

I functioned on autopilot for weeks—until Thanksgiving, when I got a call.

"What do you mean, it's not brain cancer?" I said, over and over, on the phone with my father, repeating it to make sure I was hearing correctly. I was in disbelief.

On Thanksgiving Day, I received word that there had been a misdiagnosis.

The results of the MRI had come in, and it turned out my dad didn't have brain cancer. The surgery had saved his life, but it was a vascular irregularity, not a tumor. He would live. He would be okay. I collapsed onto the couch in my new apartment and let the tears come.

The fear of losing my father had stopped me in my tracks, forcing me to look at how I was living my early adult life. I focused on being a better friend to the few people I was beginning to think of as my New York family. Daniel and Deb became my ride-or-dies; Ross, my North Star.

And when it came to plotting my career strategy, my focus was sharper. I felt freer than I had in years. Dislodging myself from anger paved the way for a more confident, fuller version of myself. A part of me opened up that I hadn't yet accessed, making room for experimentation and creativity. I was laser focused on my mission to create a startup beat covering the misfits and creatives. There was still no one at the company paying real attention to startups, and there was no manual for a production assistant to formulate a beat and run with it. For someone in my position, it was unheard-of.

I began writing articles about startups getting funded and submitted them to CNNMoney; those pieces made a bit of news, but it wasn't enough. So I continued to integrate tech into all my segments for Business Updates, finding ways to weave it into the narratives. I wrote scripts for anchors, describing apps and emerging tech trends. Ignoring the snickers about millennial self-involvement, I navigated the generational divides and became a human glossary. I briefed reporters on upcoming trends, and corrected grammar and pronunciation.

"Yes, 'tweeting' is the correct term," I said as I prepped a reporter for a live segment on Twitter, patting down her long dark hair. A producer in training, I had to make sure the talent was knowledgeable *and* presentable.

"Do people actually care about Twitter?" she inquired, looking down at the script I'd written for her.

"One hundred percent," I said with an *I know what's up* attitude. I gave her hair one final pat before she left.

As I scrolled the teleprompter, listening to her recite the words I'd written, I felt the edges of a wild thought: *Why can't I do this myself? Can't I be the one on camera?* The idea felt a bit dangerous. While I'd interviewed Biz Stone on camera, that was a one-off. In my SXSW interviews I'd spoken with the tech founders from behind Deb's cameras, producing and asking questions. At this point, I'd written and produced plenty of interviews, but those had always happened *behind* the camera. I never felt like an on-camera person. I was more comfortable directing in a lower-stakes capacity, with chipped nail polish. Would anyone take me seriously? Was I even allowed to ask for this as a minnow in the newsroom hierarchy? I'd never envisioned myself as being someone who'd appear on-air. I was more comfortable writing, and at the Desk my dream had been to eventually become a field producer. But I was evolving. The more I produced for other people, the more I wanted to write for myself and speak my own words on-screen. What had been a fleeting thought was beginning to feel more urgent. But it was still scary to admit there was a growing part of me that wanted to appear in front of the camera, so I didn't.

At the Desk, most people I knew were still cutting their teeth as news assistants. From there, the food chain included graduating from news assistant to associate producer and then on to producer. The process could take a decade. Most of us rolled our eyes anytime a news assistant expressed on-air ambitions. Every so often a producer would turn into on-air talent, but generally on-air talent at the CNN level came from other networks: people who were experienced, polished, and represented by agents who negotiated their contracts with CEOs in glass offices on the other side of the elevator banks. That's how the system worked.

But as I watched the reporter's freshly brushed hair bounce around her shoulders as she talked Twitter, I couldn't stop looking at the camera and thinking: *What if I had the courage to stand in front of it?*

Susan Grant had a cropped haircut, an infectious smile, and leather bracelets climbing up her arm. She had been tapped to head up

CNN.com before people had realized the potential of digital, before it had become an asset, and was given a home base in the Atlanta office. Although she'd been at the company since the mid-nineties, she had thankfully missed out on corporate-speak training. She was technically categorized as a "suit," but was far from it. She seemed to have absolutely no filter and the type of boldness I hoped to one day emulate when walking into a room. In other words, she was my idol, and I needed to convince her to take me under her wing.

I found Susan's address in CNN's email system, which made everybody accessible, and drafted my message, personable but professional; not too long, but enough to keep it interesting. I sent over the tech stories I'd been working on, with a few lines about the growing importance of tech.

She immediately wrote back and agreed to meet.

On one of her upcoming visits from Atlanta, we had coffee in the tenth-floor cafeteria overlooking Central Park. Susan was looking for ways to create more value for CNNMoney and CNN.com, and what I'd said about technology had struck a chord. While a part of me was convinced that I was just a lowly production assistant, another part of me felt like I had something to offer the right person who was willing to invest in me. Those people weren't easy to find at a large company, but I'd always searched for them. Often my emails went unanswered, and the rejection stung, but it hadn't stopped me. In my gut, I knew Susan was someone I wanted to build a relationship with. I craved someone who would commit to me and invest in my future, with whom I could share my aspirations.

When I met her in person for the first time, I was immediately at ease. She was quick to laugh, and it didn't take her long to give me her assessment.

"No one is doing what you're doing," she said matter-of-factly. "You've got to own this."

I was shocked at her candor and confidence. By the time we rode the elevator back to my desk filled with papers, I'd given myself permission to believe her.

Susan quickly became an ally. She was a journalism junkie and spent much of her frequent NYC trips with CNNMoney's executive editor Chris Peacock, Caleb's boss, who, like Susan, was smart, thoughtful, and sarcastic. Then she'd march through the newsroom and prop herself on my desk, pushing over my mess of papers, old scripts, and Post-it Notes.

"What's in the drawer today?" she asked, nodding at my drawer full of tech gadgets PR people were sending.

"Mr. Potato Head?" I held up the figurine, which for some reason was dressed as Darth Vader. *Thank god she has a sense of humor,* I thought. My messy, disorganized desk was becoming a repository for memorabilia sent by PR companies. My drawers were stacked with everything from electronics to figurines, even a strange Jasmine costume that an *Aladdin*-obsessed viewer had sent to one of our reporters.

Later that day, we sat upstairs in the cafeteria overlooking Central Park.

"I think it's going to be important for CNN to have a multiplatform journalist," I said, thinking about the startups I was covering as a side gig. "People are going to want someone who can produce, be on camera, and also write articles for the web. These jobs are going to come together in an interesting way."

A couple of CNNMoney writers at a nearby table lifted their eyebrows. I was explaining the future to the head honcho.

"The position doesn't exist," I continued, blocking them out. "But I think it should."

"You're onto something," she replied. "Write it up and bring it to Chris Peacock."

Could I actually write my own job position—a mix of television and digital—and then waltz into the head of CNNMoney's office to pitch it into existence? It felt like a crazy idea. Every ounce of me screamed, *You can't do this.*

"Do it," she said, reading my thoughts.

So I went back to my desk, wrote down my idea for a multiplatform tech journalist—someone who could produce and report stories on camera, and who could also write the accompanying stories for CNNMoney.com and CNN.com. I hadn't seen it anywhere else at the time, but it felt like the future. The job didn't fit into a bucket. It wasn't for a writer. Or a TV producer. It was an online hybrid. There wasn't a road map for this. I typed words over and over again, and then deleted them. *This position should have deep knowledge of tech, should be able to produce TV, have the capacity to appear on camera, and also write digital pieces.* Would Chris think I was ambitious? Out of my mind?

"Have you given the job description to Chris?" Susan asked me when she was back in town a week later.

"Not yet," I said. I'd written it, but I couldn't bring myself to follow through. Who was I to think I could be on camera?

"You're reluctant to own it. Why?"

I hesitated. "It seems like a reach, no?"

"Why? You are breaking tech news. No one else is doing what you're doing. You're creating this beat."

She was right. I wasn't even technically a member of CNNMoney, and I was beginning to break news in the tech space. Startups selling. Companies forming. I was becoming a go-to for tech news, and I didn't yet have a title. Why was I so afraid to ask for what I wanted, for what I was essentially already doing?

She leaned in, whispering, "Look around. CNNMoney is basically all men. Own it." Then she smiled and walked away. It wasn't a suggestion.

I took a deep breath and marched to Chris Peacock's office. It didn't take him long to say yes.

I left Business Updates and started at CNNMoney as the first multi-platform journalist at CNN in November of 2010, carving out what I believed was the beat of the future: startups. I was entering a new phase.

In the new position, I'd be appearing in front of the camera for the first time in an official capacity, and it turned out there was a lot that could go wrong. The list of things I didn't know was long, including: how to act natural when a camera was tilted toward me; how to focus on my interviewee instead of teeing up my next question; how not to have the camera capture me wondering if I was screwing up; what to wear. I had no idea how to dress for TV. Every pattern, color, and fit I chose to help me appear camera ready had the opposite effect. Eventually, I landed on my safest bet: a black blazer from Express.

Despite my fears and my inability to dress, I was determined.

I wanted to prove that my behind-the-scenes work finding stories, talking to sources, and writing articles could translate in front of the camera. I wanted to be taken seriously.

I called up some of my interviewees from SXSW and scheduled my first on-camera interview with Sam Altman. We were set to discuss location apps, which had started to filter from the tech scene into the mainstream. I was so nervous that I ducked into a bar on the way to the shoot to look over my questions again. I considered having a shot of whiskey, but instead asked for a glass of water. I heard Deb's words, *Fake it till you make it,* in my mind as I approached Sam in Bryant Park.

"Hi!" I said, feeling like an impostor, hoping he couldn't sense how nervous I was. If I exuded confidence, I told myself as I stood taller, no one would know.

Throughout our interview, I was terribly aware of how unnatural it felt to be in front of a camera. Sam was young and spoke quickly, but he was also smart and approachable. I was proud of my work, but when the segment landed, I played it over and over again, cringing each time.

I tried to be kind to myself. It was a start.

Luckily, the influx of venture capital and growing excitement around startups meant more companies and more crazy ideas to delve into. One of those came from Jack Dorsey, a Twitter cofounder. Jack had been fired from his role as CEO, with Ev Williams stepping in after. In a corporate game of musical chairs, he would be reinstated as the company's CEO years later, but at the time, Jack hadn't yet returned to Twitter, and the company hadn't yet gone public, turning him into a billionaire. It was 2010 and he was launching a new company—a payment platform called Square. He believed it would be a game changer. He invited me to Third Rail Coffee, a coffee shop in Greenwich Village where the owners were beginning to test out his technology.

As I walked through Washington Square Park to meet him, I wondered what he'd be like. Jack was a bit of an enigma. He hailed from the Midwest, and while he appeared buttoned up, I'd googled him and found pictures revealing blue hair from his punk rock days.

At the coffee shop, I met the CNN producer who'd been assigned to the shoot. A self-involved Brooklyn hipster, he refused to make eye contact with someone as low on the totem pole as me. Unapologetically willing me to fail, he was Deb's opposite. Up until this day, the only interaction I'd had with him was eavesdropping on his tales from his favorite joint: a performance nightclub deep in Brooklyn where naked dancers gesticulated in human cages.

Jack walked in and greeted Third Rail's staff, introducing himself. He was one of the founders of the most buzzed-about social networks, and yet he had a quiet, unassuming demeanor. He carried a small plastic square that plugged into an iPhone to accept mobile payments, a novel thought at the time.

"Hi!" I said warmly.

To break the ice, we chatted about inconsequential things. As I nodded along to the preprogrammed small talk, I mentally reviewed my questions. My thoughts were briskly interrupted by my hipster producer.

"Hey, Laurie, I'm going to need you to sit in a booster chair."

I tried to hide my horror under a too-big smile. "I'm sorry, did you say a *booster* chair?"

"Yes," he said with a smirk. "The chairs are uneven, and I need you guys sitting eye to eye."

"Just when you think you're grown up!" I said, turning to Jack, cracking a moderate-at-best joke.

He politely shifted his eyes as the producer dropped a booster chair onto my seat. I did a small jump to position my body, annoyed at the producer's efforts to undermine me in one of my first real attempts to appear as an on-camera reporter. Once we were settled, the cameras began rolling and passersby peeked through the windows. *Go time.*

"We built this little piece of hardware. It's very small," Jack said, holding up the tiny plastic square.

"It's so small, but it can do a lot of damage," I said with a laugh.

While the idea of accepting credit card payments from your phone seemed out there, anything Jack touched immediately had credibility. I was beginning to understand that if people like me thought your ideas were crazy, you could be onto something.

"It's a credit card swiper. It's self-powered. It plugs into the headphone jack of any device with audio . . . So what this allows is not just artists to sell their goods, but flight instructors, babysitters, dog walkers, coffee stores, selling something on Craigslist. We're going to give it away for free."

"How are you going to give them out? Stand on the streets?" I asked.

"That's one thing," Jack said, laughing.

I got the sense that he would stand out on the corner with a bucket of plastic squares if it meant convincing people to use his new product.

"That doesn't scale that well, though. So we're going to give them out on our website. You can sign up and put in your mailing address, and we'll send you a swiper and a sticker and a little welcome note, and you're in business."

When we finished the interview, I hopped out of the booster chair and wished Jack the best. Then I watched as he exited the coffee shop to become just another anonymous New Yorker.

I turned around, thinking that perhaps, after a decent interview, I could begin forming a bond with the producer.

"Thanks for producing," I said to the man. He grunted, packed up his tripod, and left.

I bought a latte and slipped out. A block away, in Washington Square Park, an elderly Black man with a long beard played "What a Wonderful World" on his saxophone. Tourists and NYU students paused in front of him, swaying in unison to what felt like an anthem. The notes floated through the park, which was framed against a pale blue sky. It seemed like a perfect backdrop to imagine the soon-to-be-billionaire handing out his plastic squares to strangers, while promising us a better world.

August arrived, and after weeks of cutting my teeth in front of the camera, wearing my ubiquitous black blazer, I flew to San Francisco for several interviews. As the plane landed, I couldn't stop smiling, thinking that, hopefully, this would be the first of many trips to San Francisco.

A rookie to the city, I was taken by the Golden Gate Bridge

and the billowing fog; the hilly streets that knocked the wind out of me; how the climate changed every five blocks; the brightly painted homes; the uneven hanging lanterns and tea shops of Chinatown; and the overwhelming feeling that I was entering the tech mecca where it all started. The New York tech scene was exciting, but nothing competed with San Francisco.

The next day I interviewed Twitter's CFO, Dick Costolo. Rumors were swirling in the tech blogosphere that he'd soon take over as CEO. As Twitter entered its adolescence, the big question was: How would tweets turn to profits?

"I think that the advertising platform will be one pillar of the monetization strategy," he said, rattling off other ways Twitter was experimenting with making money.

I came back to CNN with a couple of pieces on San Francisco tech and interviews with the future CEO of Twitter. Within two months, Ev Williams would step down, and Dick would take over.

Back in New York, I spoke with Instagram CEO Kevin Systrom. Before Kevin and his cofounder Mike Krieger started Instagram, they'd created an app that fell into the "location" category, a check-in app called Burbn, and had just happened to notice that people were leaving photos when they checked in. They'd paid attention to what their users were doing and completely pivoted the company.

"Our first day, we had something like twenty-five thousand sign-ups. It was a huge day for us because up until that point, everything I'd built had never gained that many users in a *year*," Kevin explained to me on camera, as we stood on the West Side Highway, cars whizzing by. The founders were in town for a conference on the West Side. Our shooting options were limited, but I figured it was worth it. The app was beginning to get traction, and I thought it was important to do a story on them while they were in New York.

Instagram was now six months old and had four employees and a little over 4 million users.

"So walk me through it," I said, moving my hair out of my face in the gusting wind. "Why exactly do I want to use this service?"

Kevin spoke like a nerdy-but-confident Boy Scout. "Instagram is a fast, beautiful way for you to share your photos with your friends." He went on to explain why the app was growing in popularity. "It really allows people to take everyday moments and turn them into really beautiful moments," he said as we snapped photos of each other. He added a filter, and although I still looked a mess, my hair flying everywhere, he was right: we both looked "more beautiful."

"How do you make this a viable business?" I asked. It was the question I put to all the social and mobile app founders who were promising to change the world with their platforms.

"At the end of the day, if we want to be a very big business, we very much are an entertainment medium," he replied. "We mix media from your friends with media from folks like CNN, ABC, NPR. We put that in front of you. If history has taught us anything, it's that advertising mediums that are the most exciting are ones where we push images to people, and that's at the crux of what Instagram's about."

I could tell they hadn't thought it out fully, but it was top of mind. The business model was quickly becoming the topic of conversation in Silicon Valley, as startups, with millions in funding, were looking for ways to make money. And "advertising" was the buzzword.

At that point, my apartment was scattered with the brightly colored business cards of startups disrupting everything from advertising to the food industry. After tech events and late nights at the tech haunt Tom and Jerry's, I'd lie awake, racing through ideas that had been relayed to me, trying to figure out which ones were worth calling attention to and which companies had larger implications.

There were a number of venture capital firms in Silicon Valley and a handful to pay attention to in New York. My phone was constantly buzzing with a new concept. I was invited to attend demo days where entrepreneurs gave their pitches: kernels of ideas that would later become apps like ClassPass. Not long after my interview with Kevin, I'd met the CEO of a company who'd excitedly explained his concept: technology that could disrupt the advertising industry. The tech enabled billboard ads to analyze everything from your gender to your age and change themselves depending on the viewer's appearance. It was newsworthy, and I released an article on CNNMoney.com. It was a two-parts creepy, one-part cool technology that could be a game changer. Apparently one of Randi Kaye's show producers thought so, too.

I was just coming into the office after a morning coffee run when I received a summons for my first live, on-camera appearance: Randi wanted me on her show in thirty minutes.

"They want me on live TV," I said, racing downstairs to Deb's desk.

"Seriously? Like *live?*" she responded, as we got up and walked down a long newsroom hall. "That's huge."

"I'm freaking out." I could feel my hands beginning to tremble. Until this moment, all of my interviews had been taped and released on CNNMoney, but none of them were live, and none of them had appeared on national television. For the first time, there would be no editing to cut down my rambling. No postproduction to save me. If I fell out of my chair or said the wrong thing, viewers around the world would see it in real time. What if I forgot what to say? What if I blanked on-air? Worst-case scenarios played out in front of me like a horror film starring myself.

"What's the worst that can happen?"

"Don't jinx me!" I gulped, rushing to the elevator, back up to Caleb's office.

I bolted through his door. "They want me live . . . on-air?"

"Time to step into the arena," he said with a smile. "Also, stop by hair and makeup," he added, eyeing the bird's nest on my head.

I raced down the hall to CNN's most lively room. There, laughter resounded as guests with big personalities prepared to go live on-air. Just a year before, I'd been a news assistant, walking guests to this well-lit room. I'd gawked at the makeup artists who were so beautiful I'd dubbed them "the Mermaids." They had long, flowing hair, winged eyeliner, impeccable style, and the ability to transform guests into their most glamorous selves by waving their magic foundation brush. Would they be confused if I sat in the chair?

"Sugar, I'm so proud of you!" Phillis exclaimed as she approached, armed with hair dryer and blush.

"Thanks!" I grinned while wondering if, based on my upcoming performance, this would be my only time in her chair.

"Go get 'em," she said, swinging the chair around, revealing an upgraded version of myself. I studied her work in the mirror, wondering if every CNN guest felt like they were experiencing prom with a side of panic attack.

While there was just enough time for hair and makeup, there was no time to change out of a shirt that made me look like I was on safari. I looked down at the orange and red print that had been a hit in artsy dive bars. *Really, Laurie? Today had to be the day you wore it to work?*

Before I knew it, I was in the newsroom, right where I used to mic up guests. On my way down, I walked by a young producer and friend named Derek Dodge, who waved. He'd been up all night tweeting news of Osama bin Laden's death from the CNN Twitter account. Like me, he was an early adopter of social media and had helped transition the company into using platforms like Facebook and Twitter. At the moment, he had quite a bit of control of the company's social media accounts. I wondered if I should ask him to tweet out my upcoming segment, but made a conscious effort to focus and do everything in my power not to become an embarrassing YouTube video.

"With the death of Osama bin Laden, the U.S. government is moving to the next objective, gaining more intelligence about al Qaeda . . ."

Sitting at camera 412—the same camera that Deb and I had schemed behind—I listened to Randi Kaye's voice through the IFB (interruptible foldback), a device you put in your ear that allows you to hear the show. As I gazed forward into the camera, I wondered how many producers whom I'd once rolled teleprompter and logged tape for were in shock that I was sitting where the guests sat. I half expected one of them to run over and walk me off set, yelling, "There must be a mistake! She doesn't belong here!"

I reminded myself to breathe.

"Meanwhile, we're going to break away just for a moment to examine something that could literally change the face of advertising: Ads that look back at you. Ads that analyze your face, your age, and sex, to change and tailor themselves to you," I could hear Randi's voice through the IFB, teeing up the segment.

"Hey, Laurie, two minutes in break, and we're coming to you," another voice spoke in my ear.

I smiled into the camera. "Great!" *Could they smell my sweat?*

The segment lasted only a couple minutes. I explained the technology: facial recognition software that sat on advertisements and determined your age, your gender, and how long you'd been looking at the ad. The ad could even change based on weather conditions. Walk by a billboard on a cold day, and you might see an ad for a hot cup of coffee.

"And what about the argument, which I'm sure a lot of people watching right now are probably arguing about at home, that these ads invade your privacy?" Randi inquired from her studio in Atlanta.

"Yes, you know, with this technology, there's always that creepy factor. You know, that stalk you," I tried less than eloquently to explain while waving my hands, as if conducting an invisible orchestra.

"Well, Laurie, I think it is a little freaky, but it's fascinating, so I'm glad we had this discussion," she said.

"It is. I know. They want the ads to be more relevant, I think that is what it is, at the end of the day," I replied.

"Sure. That's what they say," she said, before wrapping me.

And just like that, my first live segment, or "hit," was done. I breathed a sigh of relief. *I think I did all right! Caleb will be proud,* I silently hoped, as I climbed the stairs to the fifth floor and walked over to his office. He was smiling as I entered, and he held up eight fingers.

"Eight," he said.

"What's eight?" I asked.

"The number of times you said 'ya know' within four minutes."

I was mortified. My cheeks were still hot when I sat down at my desk and pulled out my cell phone.

It buzzed with an email from Ross: *I'm so proud of you!*

I smiled and looked up to see Deb coming toward me.

"Everyone at the Desk was in shock."

"I mean, I'm still in shock," I said.

"It's just the beginning," Deb said, grinning.

I felt like I was finally where I'd wanted to be since arriving in New York City three years earlier. I was getting airtime, and my days and evenings were a steady stream of tech stories and conferences and VC events. Nights off, I spent locally at Centosette with Deb or Daniel, or both.

By December 2010, Foursquare had turned down Yahoo's offer to buy the company for more than $100 million and was showing no signs of slowing down. And then it did what any startup that turned down huge cash flow would do: it threw a holiday party. The party included a who's who of New York's "new media" scene, and was held in the East Village, the favorite neighborhood of cofounder Dennis Crowley, who'd ditched the Foursquare-logo T-shirt for a gray suit. Attending were venture capitalists who held the keys to the kingdom, founders with growing cult followings, and engineers who took themselves too seriously. We all entered through an alleyway that led us down into an underground bar with exposed brick. There was an open bar and an ongoing dance party. Someone held a cardboard cutout of Dennis's cofounder, Naveen Selvadurai, and posed for photos for Guest of a Guest, a website capturing New York's young elite.

I attended as I always did—by myself. But as soon as I stepped inside, I saw a dozen familiar faces. I was getting to know the inner circle: bloggers, writers, founders—all with different goals but similar interests. Nick Denton, the founder of the Silicon Valley gossip blog Gawker, slid by me, making his way to the bar. Rachel Sklar, a figure in New York media, danced in a pale yellow T-shirt under a disco ball. I grabbed a glass of champagne, observed the dance floor, the cardboard cutout, the bloggers and founders, and took a large sip of bubbles. And then, across the room, I saw Mike.

Towering over the others, Mike had messy dark hair and a beer in his hand. He had just cofounded a company called Scout, an app that helped people connect with their friends, which was beginning to get attention. We'd been in the same room before, but had never formally met. He walked over, and we started talking and quickly discovered that we were two outsiders who were lucky enough to have found themselves on the inside.

We stood close to each other in the sweaty, cave-like vacuum, cracking jokes about the scene, watching like spectators. As he brushed a strand of long, dark hair from his eyes, he commented on the social experiment in front of us—the bloggers, the nerds-turned-kings-and-queens, the photographer who snapped photos that would end up in a New York–centric blog. We edged to the corner of the room. I didn't know much about Mike, but I immediately felt at ease with him.

"Want to grab a drink sometime?" he asked as the party neared an end.

He was a founder; I was a journalist. I had never crossed that line. I had strict rules: don't mix work and pleasure. But there was something about him that made me bend.

"Sure," I said, passing him my number before leaving the party.

On our first date, we scrolled through Instagram together, laughing at each other's accounts. He quieted when he saw the first photo I'd posted: a shot of my father. He pointed to my dad's head, where the shunt was visible.

"My dad has one, too," he said softly.

We spoke about our parents, their divorces, and our own struggles with them. Our mutual daddy issues made us feel at home with each other, and our dreams and ambitions neatly aligned.

I was building CNN's startup beat, and he was becoming king of the startup trend. I was at the forefront of tech-centric news, covering companies flush with venture capital, and his app was becoming one of those companies. Scout was growing in users and profile, thousands of people were beginning to download it, and investors were pouring in money. No one handed Mike anything. He was an engineer who would stay up all night writing code to solve complicated problems, but he loved music as much as code and hadn't been afraid to follow jam bands like Phish across the country before his day job prevented it. As I had, Mike had risen in the ranks, working long days that stretched into evenings that ended at a dive bar. He grew his team. I grew my profile. We were creative and ambitious, and if we could only navigate the demons in our heads, we could put together the puzzle in just the right way.

We started dating, but given my professional anxieties, I didn't want anyone to know. I hadn't worked this hard to become the clichéd journalist dating a founder, so we kept it secret. During the daytime we worked, and at night, we danced. Our evenings were filled with music: Edward Sharpe and the Magnetic Zeros' "Home," No Doubt's "Don't Speak." We twirled in empty bars, singing the Offspring's "Self Esteem" and laughing until it hurt. We stayed out late to deal with the intensity of our day jobs.

Through our adventures, we developed a great friendship, rooted in a mutual bond over painful childhoods that fueled our ambition as we built our careers in a boom. We shared a deep desire to be loved, and a fear that we were unlovable. We had a similar intensity that allowed us to stay up late and get up early—that helped us build and navigate the politics of media and tech while battling our own heads. We were our own worst enemies, often having destructive shouting matches in the evenings after one too many drinks at a dive bar and making up shortly after. During dinners and drinks, we melted into each other, but when we attended events, you'd never know we went home together.

Parties were my secret weapon, and I couldn't compromise my status. Venture capitalists slipped secrets over cocktails, and gossip flowed as freely as the drinks. Talk of deals was inevitable after they'd had a couple glasses of wine.

"I heard the founders just signed the papers," one VC said to me at a fancy party overlooking Central Park one Friday. "TweetDeck is selling to Twitter for forty million."

A few calls later, I learned that the deal would be announced on Monday. I slipped out of the event and raced to the office to write up the news, publishing it and watching as blogs in the tech space picked it up around the internet. It was like a sport. We were all playing the game. It was exciting—everyone was young and a part of the future.

◆ ◆ ◆

In March 2011, Mike's app was crowned the winner of SXSW. He started getting offers from companies looking to buy Scout. In the meantime, I was interviewing one up-and-coming CEO after another: from Bump, Pinterest, Shazam, Angry Birds. I started appearing more regularly on television in brief segments where I'd explain the growing influence of apps to help people shop, get a deal, or meet up with their friends. I developed a niche, uncovering startups before they were buzzy, and became a go-to journalist for up-and-coming companies to get on the radar. I started gaining a bit of traction and became a familiar face associated with the burgeoning tech wave. Flush with the promise of success, I decided to move from the East Village to the West Village. Whereas the East Village was younger, a bit grungier, and dotted with tattoo parlors and dive bars, the West Village promised cobblestone streets with names like Perry and Jane. Moving west, metaphorically, meant moving into a different, more "grown-up" phase of my city existence. I was ready to make the leap.

I looked at dozens of apartments, convinced that when I stepped into the right one, I'd *know*.

"It's like a relationship," I said to my second broker, after my first broker left me, skeptical that I'd ever settle down.

"I'll just know," I exclaimed after my thirteenth viewing, pretending I didn't catch Broker Number Two rolling his eyes.

My search for the perfect residence became a metaphor for my love life. I was waiting for "the feeling," but I hadn't had it yet. Not with an apartment, or, I was realizing, with Mike. I adored him, but was I in love with him? The question mark remained, no matter how much I tried to push it aside. I loved our matching eccentric energy, our feelings for each other's pasts and mutual support for our future ambitions, but if I let myself slow down, I wondered if either of us was the other's Boardwalk. Was it possible to love someone who was broken in the same places? No matter how much care we had for each other, Boardwalk wasn't a mirror.

Over cheese fries and gin, Daniel poked fun at me—maybe "the feeling" didn't exist? But when I walked into 26 Perry Street, it struck me like a cliché.

"This is it," I whispered, three flights up, standing in my dream apartment, looking out white-iron windows that opened onto the most beautiful block in the West Village. "I'll take it."

Shortly after I moved my belongings into my new place, I flew to London for a solo vacation. Mike was stuck at work, and I wanted to revisit my old haunts from my time abroad. I was making more money, and I'd been saving for months. I was ready for a break and a bit of escapism.

"Miss Segall, for what, exactly, did you come to the UK?" the elderly woman at customs inquired.

"Vacation!"

She eyed me. "Where are you staying?"

"At an Airbnb."

She scowled. "What is that?"

"It's like mi casa es su casa!" *Oh my god, I'm digging my own grave.* "People rent their homes or an extra room to people like me," I added, hoping to not be taken to a windowless room for questioning.

"That sounds horrid," she said, stamping my passport.

During my semester abroad in college, I had fallen in love with London's history, culture, and people. It was the perfect place for me as I navigated a quarter-life crisis, wondering, *Now that I have what I wanted, is this actually what I want?*

I'd always envisioned living in another country. I'd promised myself, when I'd tearfully left London, that I'd be back. Just seven hours away from New York, there were tiny cafés filled with dozens of British tabloids shouting headlines. I'd take them in over two demitasses of espresso, near regal parks where large swans stretched their long white necks above green ponds.

Will I ever not long to live in this city? I looked outside at the black cabs that seemed to float by and the trolleys that wound their way through the charming streets.

London was exactly as I remembered it. I looked up at the billowing clouds and drizzling rain, recalling my takeaway from my last visit: the grayness served as a shield for that bright blue sky trying to emerge—like a constant promise that the best was yet to come.

My second day, while I was strolling past the statues in Hyde Park, I got a call from Mike.

"They met our number," he said.

Skype had offered to buy Scout for $85 million. I didn't know what to say.

We both knew he would sell the company. He would become rich. His life would forever change; our lives would forever change. All the noes he'd received early on, all the late nights, the hustling, the questioning: there was a happy ending ahead.

I felt both loving and empty.

I hung up and wandered to a fancy restaurant inside a hotel filled with leather chairs and fresh-cut flowers. A woman brought me coffee and orange juice and a gleaming silver dish filled with sugar, and placed a *Sunday Times* in front of me.

"Please enjoy," she said, smiling politely.

So this is what it would feel like to be rich—mahogany lamps and people that treat you with routine kindness. As I sat at the hotel, soaking in Scout's news, I thought about Mike, and the lingering dots that existed in our relationship. We were so similar, so compatible. Why wasn't I all in? I couldn't be happier for him, but my feelings for him were complicated. This was certainly a life-changing event for his future, but a part of me hoped I would've felt in some way that it was a monumental event for *our* future. After all, if we were going to spend our lives together, this success would impact both of us. But if I slowed down, I wasn't sure we were right for each other for the rest of our lives. I pushed those thoughts aside as I touched back down in New York, doing my best to play the supporting role to someone whose life was about to transform.

When I returned to New York, Scout's sale went through. The night Mike signed the papers, I went with him and his coworkers to a crappy karaoke bar in the Flatiron District, where we sang Barenaked Ladies' "If I Had $1000000." I belted out the lyrics as the night turned into a bubbly haze, with young engineers shouting Bon Jovi into microphones in a dimly lit room after too much cheap beer.

And then I went and dropped my phone in the toilet.

I caught a glimpse of myself in the fluorescent bathroom light, my damp hair curling around my face. *Don't read this as a bad omen,* I told myself, staring back at my reflection before returning to the group of newly minted millionaires.

CHAPTER 6

The Pleasure of Business

Over the next five months, the parties grew fancier and the nights longer. The tech scene was filled with invitation-only dinners hosted by VC firms in secret rooms of expensive New York restaurants.

Adjusting to his "new money" status, Mike was beginning the search for a new apartment—one more fitting, now that he'd made a fortune. I joined him on the hunt, walking around a light-filled apartment in the Flatiron District with so many windows I wondered if the rest of the street was included in the multimillion-dollar price tag.

"What do you think?" he asked, his brow furrowed.

Am I crazy to say I hate it?

"It's different!" I said, forcing a smile.

In February, we had gone to Miami and stayed at a ritzy hotel in South Beach full of mirrors and shiny white everything. White marble columns in the lobby, white walls in the rooms, white statues.

Every day, the lukewarm pool was filled with models and fat men smoking cigars, swaying to rap music.

During the days, I curled up trying to read *The Hunger Games*, shielding myself from the pool revelers with a towel. Every night we went out for $300 dinners. I didn't understand how it was possible to spend so much on food. Everything tasted the same.

Am I being ungrateful? Impolite? Shouldn't I want this? I thought, glancing at myself in one of the ubiquitous mirrors. But all I saw was a disheveled girl with a lopsided bun and a fear of settling.

On our last day in Miami, stuffed from the too many six-course dinners, I reached a breaking point when one of the thong-clad models started twerking next to the pool.

I needed a walk. As I made my way down the winding beach path, my phone interrupted my thoughts.

"Honey, how are you?" my mom asked. I knew she could sense something was off. Our calls had become less frequent over the last weeks. The less I had to explain myself to the people I loved, the better.

I couldn't find the words to explain my jumble of thoughts. All I could say was "Mom, I just want a taco."

Mike and I both attended the 2011 SXSW. He was staying at the Four Seasons, and I was staying at the Driskill, a historic Austin hotel. The days of sleeping in a double bed with Deb were over. Deb had also moved on. She left the Desk after getting her dream job at *The Next List*, a documentary show that let her travel the world and tell stories of innovators across the globe.

Companies now had their own Instagram and Twitter accounts, and departments dedicated to social media were springing into existence. CNN got the memo and sent me with a small crew; they also planned to launch the CNN Grill, a fancy, invitation-only pop-up where the network hosted events throughout the conference.

It was the year of group-messaging-app wars, and the festival was buzzing. A handful of apps that allowed users to text troves of friends at once had launched, but there had yet to be a clear leader in the space.

One of my first interviews was with the founder of an app that was gaining traction: Uber. I had connected with him through a friend a few months before.

Hey Laurie, she wrote. *I'd like to introduce you to Travis of Uber. Uber is an amazing, geeky car service that's shaking up SF & Silicon Valley . . . Travis: Laurie writes about tech for CNN Money . . . So, Laurie, Uber. Uber, Laurie.*

The idea sounded absurd at the time: get into a stranger's car for a ride. Everyone in my non-tech circles seemed slightly horrified when I ran it by them.

Travis Kalanick showed up at the CNN Grill for our interview wearing a cowboy hat and a tan button-up shirt. If companies were physical manifestations of their founders, the shoe didn't fit. Travis wasn't geeky.

Wearing my ever-present black blazer, I looked into the camera for our taped segment.

"We're sitting here with Travis, the cofounder and CEO of a company called Uber. For those of us who don't know what Uber is, tell us what you guys do."

"Our motto says it all: We're everyone's private driver," Travis said. "You push a button and in five minutes, a town car rolls up, opens the door for you, and you're on your way."

Uber had just experienced its first backlash when it raised prices during the New Year's Eve rush, so I asked him about the controversy.

"You guys recently, over New Year's, tested out 'surge pricing' and you had a massive backlash. Are you going to do this again?" I asked.

"I wouldn't say a *massive* backlash," he replied defensively. "I would say there were a number of folks who took a ride on New Year's Eve," he continued, his voice getting louder. "They were told the price at which it was going to be, but maybe they had a few drinks, or they're not used to having a dynamic pricing mechanism around personal transportation."

Interesting that he's blaming the outrage on drunk people, I thought.

We wrapped the segment, and I watched him walk off.

Travis was the kind of founder who tested boundaries. He didn't seem like the typical Silicon Valley entrepreneur; he felt more like an outsider, shoving himself to the front of the line and daring anyone to tell him to move back. It was either a recipe for a wildly lucrative and transformational startup that could upend a traditional industry or a slow-motion train wreck. The outcome was yet to be determined.

That same day, I interviewed the founders of Instagram. Since the year before, when I'd talked with Kevin Systrom on the West Side Highway, the company had grown from 4 million users and 4 employees to 30 million users and 13 employees. It was now incredibly valuable, and the tech blogs were buzzing. Would the founders sell? We made our way downstairs, again to the CNN Grill.

"Any chance that you guys would sell, or do you feel pressure to? Have you gotten offers to hand over this technology to a bigger company?" I asked.

"We're pretty focused on remaining an independent company right now," Kevin said.

Kevin was entering a new phase I dubbed "startup diplomat," which consisted of founders mastering the art of not saying anything. The stakes were rising, and the media were starting to ask harder questions.

"It really excites us to come into work every day and be able to work on Instagram," he added, as if accepting an award. "And I think that's going to be the case going forward as well." He continued to respond as if negotiating a peace treaty, as I pried about selling the company.

"I think they'll sell," I said to Mike that evening after a long night. We lay in bed, exhausted, recounting our days. It was 3:00 A.M., and we could still hear music spilling in from the streets and laughter from late-night goers stumbling into the lobby. I was prepared to get up early, rush out, and go live with my segment outlining the trends in tech on display that year.

Days later, Facebook bought Instagram for $1 billion. The promise of growth and potential competition had been enough to spur Mark Zuckerberg to offer ten figures for the app, the biggest acquisition to date for a startup. I figured he was doing the paperwork just as Kevin had gushed about going into work every day. The company's sale sent startups into a new era.

"A billion dollars for an app. It's unprecedented!" I proclaimed on TV. It would turn out to be a steal.

The following month, Facebook went public in one of the most anticipated IPOs in history. Despite the hype, the stock fell, share prices tanked, and investors were livid. I flew to San Francisco to hear Mark Zuckerberg speak for the first time since the big dip, at a who's-who conference called TechCrunch Disrupt, run every year by the tech blog TechCrunch. Startup founders attended the conference to get in front of VCs and bloggers, and the ones who'd "made it" pontificated onstage.

People stood outside the building, lanyards around their necks, startup names across their shirts, handing out business cards and reciting their one-liners to anyone who would listen. Inside, there were rows of booths where entrepreneurs pitched their concepts, promising to disrupt industries and transform society with lines of code. Then there was the auditorium, filled with people who hung on the words of the successful founders who'd come before them. It was almost a rite of passage. Even *Beavis and Butt-Head* creator Mike Judge attended, to do research for the sitcom *Silicon Valley.*

This year, all anyone could talk about was Facebook's falling stock price. People wondered if social networking *really* had the legs to stand in the public market. As Wall Street reacted negatively, there were more and more naysayers.

Attendees were packed inside, waiting for Mark Zuckerberg, a billionaire at twenty-eight, to take the stage. He rarely did appearances, especially at a moment when his company was being questioned. Bloggers lined the aisles. I could barely get to my seat, two rows from the front. We waited, and the market waited. Twitter blew up with anticipation.

As bloggers tweeted and the auditorium buzzed, I caught a glimpse of Mark in the corner, behind the main stage. He breathed deeply, lifting his shoulders like he was getting pumped up for a ball game. It was something I might do before a live television segment. The gesture made him seem youthful, vulnerable. Then the guy who made "Facebook stalking" a thing walked onto the stage in jeans and a brown T-shirt.

"Okay, you ready?" asked Michael Arrington, executive editor of TechCrunch. A notorious Silicon Valley figure, Arrington was known for sharp questions and no bullshit.

Mark Zuckerberg's face turned slightly pink, and he tensed in his chair.

"So what's up with the falling stock price?" Arrington asked.

Mark spoke on message, saying that the company was quickly moving toward mobile. At the time, increasingly fewer people were accessing websites on their computers—they were switching to their smartphones and using mobile apps to consume information. And Facebook, which had not been a "mobile first" company, was behind. People who accessed the site on their iPhones complained that it took too long to load, or often crashed. Overall, it was a buggy experience, and Facebook was losing users. But not to worry—Mark promised he'd made mobile the social network's main priority, with an all-hands-on-deck approach and a company-wide pivot to address the problem. Buying Instagram, an app that was designed specifically for the mobile phone, was only part of the solution. The company had to go fully mobile, and its investors realized it.

Zuckerberg needed to instill confidence that the company was making the right moves, but as he described the new plan to Arrington, his words jumbled together, as if he were a student explaining to a teacher why the homework assignment wasn't complete. From my seat, just rows away, I picked up on something: Zuckerberg was as young as I was, and under the bright lights, in front of the public market, the bloggers, and the naysayers, he was nervous.

When I got back to my hotel, I face-planted on my bed and didn't wake up until the following day. When I turned on the television, talking heads were regurgitating Mark's words, but the most interesting piece of the interview, to me, was missing from all the coverage: the fact that puberty is painful, especially when you're prom king. As clips of the interview played on repeat, and Mark's words were dissected, I saw the guy who seemed so young, with a company so huge, pumping himself up in the wings of the auditorium to explain his missteps—why a company known for innovation was late to the game on one of the most important shifts in tech space. While the Harvard dropout was quickly becoming one of the most powerful people in the world, in that moment, he was extraordinarily human.

I called Mike that night from my room overlooking Union Square in San Francisco. It was blocks away from the Ferry Building, where I'd spend the mornings walking through the fog, waiting for it to break and the sun to rise against the Bay Bridge.

We spoke of our workdays, and before we hung up, he told me he couldn't wait for me to come home so I could continue with him on the hunt for a new apartment.

I wanted so badly for this wonderful man to be it. I wanted to deny the feeling that kept creeping up on me—that this was close, but no cigar. This wasn't Boardwalk, the spectacular that we both deserved.

As the months went on, and we rode the roller coaster that was heading to the end, it became clear that we'd become shadows of ourselves, holding on to the obvious: We were good people, we were kind people. But ultimately, we were not in love. The fights became more frequent. Subconsciously, I showed him in every way I could that I had a foot out the door as I came home late and avoided his texts. I stayed out of fear. I was afraid to jump, afraid of overwhelming loneliness and the punch-in-the-gut pain that happened when two people parted ways.

When we finally ended things, I slept for days, exhausted. On day five, I pried myself out of bed and walked over to an old coffee shop on Perry Street, where elderly Village locals gossiped and phones weren't allowed. I wrote sentences on scraps of paper and let the pain pour out of me. I was surprised that it manifested physically—a stomachache that came in waves—but I was grateful to feel something again. With music pumping through my headphones, I spent hours roaming the quaint West Village streets, making my way over to the river, and thinking about my father. Our relationship was better. We were healing, but would I ever have a shot at what I didn't have growing up—stability? Would my mother eventually find someone? Would I hold a feeling of responsibility for others—her happiness and pain, his absence—or would I learn to grow up, to move forward, to build healthy relationships, from a blueprint I'd never seen?

As I watched the cruise ships setting sail on the Hudson, I thought about how afraid I'd been to let Mike go. I'd never cared about a man so deeply or hurt so much. I'd let someone in. It felt like a start.

I could never have envisioned that the end of one relationship would mark the beginning of another.

The woman's name was Erica Fink, and when we met, she was just as skeptical of me as I was of her.

We had been set up to work with each other at the urging of my smart, no-nonsense digital editor, Stacy, who wasn't afraid to say anything to the bosses in the corner offices. I'd started working with Stacy as soon as I'd transitioned from Business Updates to CNNMoney, and she'd become a behind-the-scenes supporter of my career. While Caleb spearheaded video assignments, Stacy edited all my written pieces devoted to startups and technology. She was the type of scrappy editor you'd take to a war and would use for your "call a friend" card on a game show. Stacy liked the idea of me doing a story on Facebook's first earnings report as a public company. I would write the report, but as part of my new multi-platform status, she also thought it would be a good idea to shoot a video to capture all the businesses beginning to make money off Facebook.

Although Stacy had no particular authority in the video department, she was a respected voice in the newsroom and decided that I should be teamed up with Erica, who was being deployed across CNNMoney to help writers appear more polished on camera. As I'd predicted, multi-platform journalists were becoming a trend, and as digital merged with television, writers were in desperate need of on-air training. It seemed I'd paved the way for others looking to take on a multifaceted role.

I knew of Erica. We'd started at CNN at the same time. But while I worked the Desk, she worked nights on *American Morning*, CNN's morning show. Like me, she had started at the very bottom, chasing stories, cold-calling strangers, fielding unusual requests.

We were both twenty-five—born only two days apart—but while I was raised in Georgia, she grew up on the Upper East Side of Manhattan. She was completely sincere when, after I mentioned my southern upbringing, she asked whether I had bears in my backyard.

Erica was tall, understated, and well put together. A natural beauty with big brown eyes and long eyelashes, she rarely wore makeup, but when she did, she could have been confused for a television reporter. I'd seen her walking down the halls in the early mornings on my Desk shifts. Unlike me, she'd mastered the always-in-a-hurry-but-never-disheveled newsroom stride.

At our first meeting, Erica was very polite, but it was clear she thought I was just another bratty, entitled girl. In her defense, I did look like every other girl at summer camp who'd been mean to her. To me, Erica looked like the girls at college, next to whom I'd felt like an alien.

"Hey!" I said. "Welcome to the tech beat."

An on-camera reporter in the making, I was constantly on a journey to be liked, and I wanted to win Erica's approval.

"Hi," she said, turning her attention to me. I tried to read her expression. Nothing. She didn't smile or frown.

We tiptoed around each other. On one of our first stories we shot in the field together, we politely discussed a tech startup that was trying to cut out the middlemen and help farmers make more money. Erica, who was much more detail oriented, entered organized notes into what I would learn was her bible—"the planner." She arranged for a photojournalist to accompany us upstate to a farm where the startup was in use.

"I don't really shoot stand-ups," I said, pulling her aside from the cameraman as clucking chickens surrounded us.

"What do you mean?" she asked, studying my face.

A stand-up is where you stand, stare straight into a camera, and say something to introduce the piece. They are typical for on-camera reporters, and I was terrible at them. I constantly looked into the camera, repeating the lines I'd rehearsed in my head, but as soon as they came out, they didn't feel natural. They were jumbled and abrupt, and I'd often forget what came next. With each take, I feared that it was painfully obvious I was a rookie and wondered whether the photojournalist and the producer were judging me.

"It feels weird to look in the camera and speak. How does anyone look natural doing it?" I asked.

Erica didn't blink. "Practice," she said. "Everyone sucks at these at first. You just have to think of the camera like you're talking to me."

I took a deep breath.

"Try not to overthink it."

We walked back over to the camera.

"Look, if you screw it up, only the chickens will know," she added.

We laughed, the hens clucking around us.

As soon as our cameraman started rolling, I delivered the stand-up in one take. I saw a smile creep onto her face.

After that shoot, we bonded. I loosened her up, and she gave me structure. I brought in a string of founders turning their companies into multimillion-dollar ventures who pitched us ideas. Erica nodded intently as we took notes on everything from cloud computing to security startups.

We were a match made in heaven. Erica was interested in technology, and I was living and breathing the emerging startup world. But most important, both of us had been separately scheming, and together, we started to dream. Instead of an endless series of founder interviews, we wanted to do something bigger, and tech was our way in. Armed with Twitter, Facebook, and Instagram, Erica and I started to use tech to break news.

My experience told me that a tech platform was doing well when the bad guys started using it. So when Erica and I heard rumblings that, increasingly, pimps were using social networks to recruit women, we decided to investigate.

With the help of a woman named Andrea Powell, the executive director of FAIR Girls, an organization that helped victims of sex trafficking, we were connected with a victim in Oregon who'd been recruited by a pimp on Facebook. If we could just get Caleb to send us there, we'd have our interview.

"It's a no-go," Erica said after she spoke to Caleb, her expression crestfallen.

We didn't exactly have a huge budget at CNNMoney, and I wasn't a preferred talent who could jump on a plane at any given moment. But in our minds, no was only a couple of steps from yes.

We figured out a work-around. CNNMoney was sending Erica to CES, a tech conference in Las Vegas about gadgets.

"What if I pitched a stopover in Oregon, and just offered to shoot the interview myself on the way back? You could join in on Skype, and CNNMoney would basically pay nothing," she thought out loud.

Erica wasn't a photojournalist, but she knew enough about how to shoot. It was this or nothing.

She presented the idea to Caleb, who eventually said yes, since we'd given him a no-cost option. One week later, Erica stopped in Oregon on her layover back from CES. She met "Nina" (we changed her name for anonymity) at a hotel, I joined in virtually, and we tag-teamed the interview to get her story out. Nina looked into the camera and recounted a horror story: She'd received a friend request on Facebook from a guy she thought was cute. They spoke for a month until she left home to meet him. Then he abandoned her on the street, instructing her to "catch dates." She went from preparing to go to college to becoming sex trafficked. Looking into Erica's lens, Nina calmly described being beaten with a pistol and raped, then locked in a closet for twenty-four hours.

Weeks later, another victim we'd connected with through Facebook traveled from Washington, D.C., to our offices in New York for a face-to-face interview. Erica created a room full of dark shadows where we could hide her identity during filming. On camera, we referred to her as "Lisa." She wore a pink-and-white-striped shirt and could barely make eye contact when we greeted each other.

She'd been trafficked for much of her adult life. As we sat across from each other, she described her own recruitment on the platform, and how she still received over twenty Facebook messages a day from pimps.

Social media was spurring a new class of crime, and the implications were sitting in front of me. Lisa was vulnerable and raw. I felt protective of her, even as I asked more prying questions, still trying to remain delicate: *What did he say? What happened after you guys met?*

As I listened to Lisa describe her horror story, I felt a new sense of responsibility. After we finished speaking, she slowly removed her mic and emerged from the shadows of our setup, still struggling to make eye contact. The trauma was evident, and the more it was ignored, the angrier I became. Why didn't tech companies know the extent of the damage being caused on their platforms?

In addition to the interview, we taped Lisa sitting at my desk, pointing out still-active Facebook pages of pimps bragging about their prospects, posting pictures of wads of cash. We were in shock. Not only were the pages active, but the perpetrators weren't even trying to cover up their exploitation of women like Lisa. I knew that Erica and I were onto something.

As Lisa left our offices that day, I wondered whether she'd be all right after our long talk, recounting the details of her abuse and looking over the pages of the men who'd harmed her. It was all a delicate balance.

Instead of rushing out a story in less than twenty-four hours, as we usually did, we took weeks to interview women, who'd all asked us to hide their identities for fear of retribution. When we finally finished, we submitted the material to our editors.

Chris Peacock called us into his office.

"I just want to make sure you did everything by the book," he said.

"Of course," we replied.

Chris informed us that people above our pay grade, above *his* pay grade, had been contacted by Facebook in what we could only consider an effort to kill the story. Tech companies were getting more powerful every day, and the stakes were getting higher, too. He wanted to make sure we were bulletproof. Not only were we bulletproof, but we were also inspired.

"Wow, I've never experienced anything like that," Erica said as we walked out to gather our notes, in case we'd have to submit them.

"I think it means we're doing something right," I said.

We cut together the video into a longer segment that would appear on CNNMoney.com. If we were lucky, CNN.com would pick it up on the homepage. Erica and I wrote an article to go along with the video, hoping more people would see the piece.

But we wanted more. Even though I'd become a go-to reporter for all things tech, there was still a gap between the digital world and the on-air world. CNN TV was a different beast. My taped videos had a regular cadence on the website, but I had yet to break through on-air. Technically, I lived in the digital space, but the word "digital" was still foreign to many producers and reporters. And it was a tougher sell to convince producers—especially producers who knew me as a "prompter"—to take my digital videos on-air. But it was an important story, and Erica used every contact she had to make sure it aired. We sent separate emails to shows, made calls, and essentially offered to donate our kidneys to anyone who'd consider airing the piece.

It worked. The Facebook pimp piece was slated during one of the afternoon shows the following week, shortly after the arrival of CNN's new president, Jeff Zucker.

From the day Jeff arrived at CNN, everyone wanted to make a good impression. Reporters circled his office, hoping for a few minutes with the man who now held the keys to their career trajectory. Top producers shuffled in and out with an air of importance. Jeff brought with him a reputation as a TV maverick. He'd become executive producer of the *Today* show at twenty-six, and had become a name in the media spheres. People were excited to see how he'd shape the network, and those who'd been at the company for a long time wondered what a new commander in the corner office would mean for them.

Jeff immediately became a fixture in the newsroom, stopping to talk to reporters and writers, perching on their desks. His sharp eyes constantly glanced at the wall that displayed the news numbers of the day. His office was filled with television screens airing every news station, so he could watch the competition in real time from his desk.

He was speaking to Bureau Chief Darius Walker, my old boss from my days mic'ing up guests (who later recounted the story to me), when he looked over at one of the screens and saw me interviewing Lisa in shadow with a banner that read, "Pimps recruit women on Facebook."

"Who is that girl?" he asked.

"That's Laurie Segall," Darius replied.

The next thing I knew, I was summoned into Jeff's office. Cursing myself for not brushing my hair that morning, I patted down what felt like a nest forming while power walking into the most coveted six hundred square feet in the newsroom.

Jeff pointed to one of his screens and said, "We need more enterprising journalism like this. Not just regurgitating the news of the day."

It took a moment for his meaning to sink in: he thought my work mattered, that we needed more of it. To most of CNN, I was a sometime production assistant who'd gamed the system by becoming an on-camera "talent" of sorts. Validation from the big boss meant an automatic level up. I had been recognized by the one person who mattered. Jeff saw me not as what I had been, but how I saw myself—a journalist paying attention to technology and social media in a brand-new way. He understood that I was onto something—and that "something" was important. It was the future.

I turned around to leave, trying to hide the smile plastered on my face, when he added, "Oh, and we've got to fix your tracking voice."

I cringed. I hated voiceovers. Even while I recorded them, I knew they were coming out wrong. I wanted so desperately to be taken seriously on camera, to have gravitas, that I became a caricature of myself. I tried to speak like a female Anderson Cooper, but inevitably, I sounded like I was narrating a funeral.

I ran to Erica's desk and told her the good news, plus the dilemma. Together, we found a tracking booth, and she had me speak into the microphone over and over again, until I started sounding more natural.

The following morning, an email from Caleb landed in my inbox.

Congrats, Segall. You've broken through.

Jeff's daily 9:00 A.M. meeting took place in the conference room dubbed Strawberry Fields, where senior executives sat in order of importance to discuss stories of the day. It was there, according to Caleb, that Jeff mentioned my name for the most important show producers, editors, and bigwigs to hear: "Is Laurie Segall looking into this?"

From then on, Erica and I started each morning with an email from Caleb that looked like this:

> Monday, 9:43 A.M.
> **FROM:** Caleb Silver
> **SUBJECT:** [Insert Story of the Day.] Jeff wants you on it.

CHAPTER 7

Professional Stalkers

It was 11:00 P.M., and Erica and I were still in the newsroom, splayed on a couch in a back room, Instagram messaging and tweeting at anyone associated with Dzhokhar Tsarnaev, the person who was believed to have set off a bomb during the 2013 Boston Marathon.

Earlier that day, the media had identified Tsarnaev's Twitter account and profile on VK, a Russian version of Facebook. A producer who was on the ground in Boston, talking to people who knew Tsarnaev, sent us an email: *FYI . . . hearing Tsarnaev was on Instagram, "jmaister1" may have been his name.* It was the perfect tip to send us down an internet rabbit hole.

We were quickly becoming super-stalkers. We understood technology, had access to the people who created it, and used the internet to creep on people for reporting purposes.

"It says his profile is deleted," Erica said, sitting cross-legged next to me on the ugly green couch.

"Okay, but we know that on the internet nothing ever disappears."

We typed "@jmaister1" into Google. Although we were met with "User not found," we clicked on the link anyway, and a dropdown menu offered up an arrow that said, "Cached." Google's web cache stored information for about a month after it was deleted.

Sure enough, in a couple of clicks, we were able to see a shadow of "@jmaister1." Using cached websites, we pieced together Dzhokhar's account and found jihadist pictures and hints of extremism: digital traces left behind by the man who likely shook the heart of Boston on a day that should have been a celebration. We could also identify the usernames of friends who'd commented on the pictures.

I had his deleted images, clues to help us understand more about him, but one thing stood in the way of going on-air with it: we needed people who knew him to confirm that this had been his Instagram handle. Two sources would be enough for "the Row"—CNN's most seasoned editors, who sat in a row in the Atlanta newsroom, picking apart pieces before they could air—to confirm the story was editorially bulletproof, and to give our news package the go-ahead.

Into the night, we sat in the fifth-floor newsroom, on Instagram, @messaging his friends—who appeared to be a mix of pot dealers and druggies—to confirm our story.

Finally my phone rang. I'd been waiting on the call.

"This is Laurie," I answered a little too enthusiastically.

Bingo. It was a friend of the probable Boston bomber. I had tweeted him and messaged him my cell phone number. All I needed was for him to confirm that this Instagram account belonged to his friend Dzhokhar Tsarnaev.

"Who are you?" an angry man yelled into the phone. Before I could identify myself, he interrupted. "I don't like what the media is saying!" he screamed as I paced back and forth.

"Hey, I'm not looking to speculate," I responded, trying to lower the temperature. "I'm just looking for some info on your friend's account. We're just trying to get a more accurate picture."

The guy breathed heavily on the other end of the line. This was the moment: either he hung up or we transitioned to a tense but civil conversation. My instincts pushed me to keep talking. *Don't let him decide.* Before he could make up his mind, I kept going.

"Look, I know all of this is surreal," I said as Erica listened, her brown eyes wide, scribbling notes. I had scoured his Instagram profile for a way in, something personal and relatable, but mostly, I'd gathered that he smoked a lot of pot.

"I don't believe he did it," the friend said defiantly.

"I hear you. It's not why I asked you to call," I replied, softening my tone, trying to envision what it would be like to learn that your friend was a mastermind behind a bombing that killed three people and injured at least 264 others.

"Honestly, I can imagine it's a lot," I tried to empathize. "We're just trying to figure out if @jmaister1 was his Instagram account," I said, teeing up the question that would get us to air.

He paused. Erica and I looked at each other. I pressed *speaker* so she could listen in. I could see her knuckles turning white.

"Yeah, that's him," he said, his voice quieter now. "He deleted it recently."

Confirmed.

"Hey, thanks. I know this is a strange time."

The man hung up.

"Amazing, but not enough for TV," Erica said. "We've got to explain how we resurrected this thing. No one understands the internet. Do we have someone we can put on camera to explain Google's web cache?"

Erica and I still had to send our story to the legal department and explain how we were able to re-create an Instagram account that had been deleted. We needed a tech expert.

"Sam Altman is coming to the office for coffee, first thing tomorrow," I said.

Sam and I had planned to have a catch-up. It had been a while since we'd sat across from each other for my first official on-camera appearance, and he'd become my go-to for all things Silicon Valley. At this point, Loopt had shut down and been acquired by Green Dot Corporation, and Sam had become a part-time partner at Silicon Valley's most prominent incubator, Y Combinator, which had invested in Loopt, and was known for investing in companies like Airbnb and Reddit.

"Think Sam would go on to explain how tech can be used to investigate a bomb plot?" Erica asked.

I thought about Sam, the whiz kid, who was the walking embodiment of anything and everything representing technology. He was quickly becoming a Silicon Valley brand name through his thoughtful blog on company building. I was sure he'd be down for it.

"I have no doubt."

The next day, Erica was directing a camera over my shoulder as Sam pointed to traces of Tsarnaev's deleted Instagram profile on Google's web cache.

"It's exactly like an archive," he explained as he sat at my computer. "So no matter what changes were made to the page today, on the current server, Google has this sort of imprint from a couple of weeks ago."

We showed the digital traces that Tsarnaev had left behind: he'd liked several photos referring to Chechnya that were posted by other Instagram users, including a Chechen warlord, Shamil Basayev, a onetime government official who later masterminded terrorist attacks against Russia. Sam's show-and-tell helped put into context how we were able to find out information.

I thanked him and said, as I walked him to the elevator, "Next time, more Silicon Valley, less devastating terror attack."

Just as the elevator doors shut, my phone rang. It was someone from the FBI, inquiring about our report. We'd reached out to law enforcement for comment, and now they were asking us for information.

"Is this Laurie Segall?" an official on the other end of the line said formally, introducing himself.

"Hey there. Yes. Did you guys have a comment for us?" I responded, hoping to get some formal statement from law enforcement on the record.

"I was hoping you could share with us any more information on this so-called deleted Instagram account."

I was surprised. It was an odd request. Law enforcement didn't normally make calls like this. The proper protocol was to see a completed story, not ask for our help in their own investigation.

"We are going live shortly," I responded. "It will be public soon. Let us know if you guys have a comment."

My heart was beating fast. We needed to get our story to air.

"We've got to talk to legal ASAP," I said to Erica after hanging up. This story was a far cry from covering startups.

After we'd spoken to our legal department and cleared the images to use, double-checked our sources, and ensured that the story was rock solid, we were ready to take it to air.

With Jeff Zucker on my side, it had become a bit easier to get our pieces on television, but Erica and I were still used to receiving noes—ranging from polite to dismissive. But we never gave up. Instead, we developed a ritual. Every time we felt strongly about a piece that was rejected from a show, we dropped everything and played what we'd deemed "our song." From my iPhone, we blasted Janis Joplin's "Piece of My Heart," refusing to let the rejection sting. Of course it did, but something about Joplin's folksy voice shouting into the newsroom made us feel like crusaders.

"Segall, I think this story is prime-time worthy," Erica said.

I could see excitement in her eyes. She was right. We were breaking news and uncovering something that hadn't been out there. *Why shouldn't we pitch Erin Burnett or Anderson Cooper?* Their shows were the gold standard, and while I was nervous, I knew this story had major legs. I stood over Erica as she carefully drafted the email to Erin Burnett's executive producer, a friendly, no-nonsense man who might give me my big break on prime time. I hoped he'd forget that he'd once referred to me as "prompter" and could now see me as an investigative reporter.

"And . . . send!" she said, looking over proudly.

Within minutes, he got back to us.

Let me see if I can fit this in tonight. Will know in an hour.

Erica and I were ecstatic. We would be on prime time! Erica wasn't a hugger, but I could see her excitement. She scribbled into her planner and tapped her foot under her desk. I vaguely made out a smile. We both turned to our computers, waiting for the email.

An hour later, he wrote back. *Sorry, the show is jammed tonight. Good story, though.*

I looked over at Erica. Both of us were crestfallen. "Secret room?" I asked.

The tiny space was reserved for guests appearing in the fifth-floor studio, but the studio was rarely in use, so it became our secret hideout. Amid an open newsroom, this little slice of privacy offered a beautiful view of Central Park, and most important, what Erica and I dubbed "the therapy couch," where we discussed frustrations over coworkers, managers, and boyfriends.

We barely made it down the long halls, by the control rooms, before I pressed *play* and Janis Joplin's voice rang out.

"*And each time I tell myself that I, well I think I've had enough,*" we sang along, twirling around. "*But I'm gonna show you, baby, that a woman can be tough.*"

We published the story online, and the next day, Wolf Blitzer's *The Situation Room* invited me on to discuss it. As Erica and I walked back to the studio, I looked over my notes, thinking about the beats I'd hit on-air. She made her way to the control room as I positioned myself in Studio 53, plugging my IFB into my ear and listening to the familiar sound of CNN programming.

"You got me, Segall?" I heard her voice.

"Yep." I looked at the screen in front of me, which was airing the show I'd be appearing on. Images of devastation sat behind a chyron that read: "Exclusive: Tsarnaev's deleted Instagram."

"They're teasing your piece now," she relayed. "A minute until you're on."

"Great." I took a deep breath and relayed our findings live on-air.

As I glanced at my notes, ticking through my points, I thought about the startup beat I'd helmed and the emerging tech that helped us carefully craft a fuller portrait of the man responsible for the devastation of that day. I'd made my way on-air through a side door, and this one felt more interesting, and increasingly relevant.

And while we hadn't managed to get our piece on prime time, it still made it to air and broke news on an important story. We called it a win.

Erica and I had figured out what other newsrooms would soon pick up on: tools like Instagram, Facebook, and Twitter weren't just changing our social culture; they were transforming how the media did their job.

We were hitting our stride, together, investigating national news through social media. As our lives intertwined, Erica and I became closer. She became the person who finished my sentences, the one I spent late nights and early mornings with—before she went home to her fiancé, Arel, a brilliant engineer. Seeing Erica and Arel together, so perfectly matched, made me wonder if I would ever find my own perfect match. Would I care about someone as much as I cared about the stories I was covering? Would my empathy for these subjects translate to a relationship? Could I let myself slow down from the adrenaline rush of the newsroom, from the sprints of daily stories? I missed Mike's solid nature, his all-encompassing hugs, and the feeling of safety that came along with a relationship.

While I wondered how he was doing, Mike and I didn't keep in touch. It was too painful to pretend we were friends. There was resentment because our relationship had stretched on too long, with me lacking the courage to say out loud, "I have to go." But freedom from it had done me good. I was moving toward a version of myself that I was proud of: someone who was working hard to make a name for herself in a newsroom. I was carving out a beat and building a support structure filled with people like Susan, who'd encouraged me to write my own position that had previously never existed; Stacy, who continued to edit my stories and push me to grow as a journalist; and Erica, who always had my back as the days got longer and the stories more intense.

And I could see the results. Less than a month after Erica and I pieced together the deleted Instagram account of the Boston Bomber, I landed the biggest scoop of my career, and got my shot at prime time.

It all began when a woman escaped from a basement in Cleveland, Ohio, on May 6, 2013. She and two other women had been kidnapped and held hostage, for a decade, by a man named Ariel Castro. When she bravely broke free and ran to safety, the mystery of three missing women was solved. Castro was taken into custody, and the other women were liberated. The story went national.

ABC, NBC, and CNN flew to Cleveland, microphones stretched toward anyone who would speak. Everyone wanted the exclusive with Castro's friends or family—no one was off-limits.

At 9:43 A.M. that Tuesday, Erica and I got the email from Caleb:

JZ wants you to look into Castro's social.

On it, I responded, and went to work digging into Castro's digital footprint per the request of Jeff Zucker.

Within minutes, I had found Ariel Castro on Facebook, and started sending friend requests to all of his friends. I was familiar with Facebook's then-loose privacy settings. I knew that if any of

Castro's friends became my friends, I could gain more access to his profile. One of them did, and I was able to see that the same woman kept posting on Castro's wall. Her name was Angie Gregg. With a few clicks, I figured out that she was Castro's daughter. Clicking the button to add her as a friend, I then sent her a message—*I know you're going through a tough time*—and gave her my cell phone number. Maybe she'd get in touch.

I was growing used to reaching out to people in the moments they became strangers to their own lives. I had empathy for them and was always careful to be respectful of their pain. But I also was asking something of them, and so I walked a fine line between reaching out and pushing too hard, especially as reporters from other networks also reached out. The only way to break through was to develop a connection.

But I knew early on that reaching out to people in their most vulnerable moments wasn't a sport, and getting them to agree to come on camera wasn't a win. It had to be more than that. Their story had to move the needle and have impact. Otherwise we were but vultures preying on a moment.

Later that night, I was in a crowded bar, enduring painful small talk with a guy I had met online. My phone buzzed, lighting up with a Cleveland area code.

"I've got to take this," I said, throwing a twenty on the bar and racing out to the dark street. On the other end of the line was Ariel Castro's mother, yelling at me. I assumed she'd received my number from Angie Gregg.

"You've got to take them down!" she said, in between sobs. She was angry and understandably hysterical. I peeked back into the fogged windows of the bar, where my date had started flirting with the bartender. I didn't feel bad as I walked toward Perry Street, and focused on Castro's mother.

She was furious with CNN. The network had been airing mug

shots of Castro's brothers, who'd been apprehended but were found to be innocent. They were no longer suspects, but their pictures were still appearing on-air.

"I hear you," I said as I thought about her, sitting in Cleveland, staring at the TV screen, her world crumbling. I told her I'd call the control room in Atlanta, and immediately hung up and dialed the number I'd memorized as a news assistant.

As soon as we stopped airing the pictures, I called her back.

"They've been taken down," I said. I wanted to ask her more. How was she doing? What did she know? Were there ever signs that her son had the ability to create a real-life horror film in his basement? But building trust with a source is tricky. You have to know when to push and when to pull back, especially when you walk into someone's nightmare.

I made the decision not to push. I could tell she was, justifiably, emotional. So much was out of her control at this point. Her priority was to grasp what she *did* have: two sons that were not part of the horrific story that would now become her life. I'd helped her with that, and chances were, I could ask for more. But I wanted to give her a moment to rest.

We hung up, and I shot a text to my date. *Great meeting you tonight! So sorry I had to jump. Looks like things are crazy right now. Good luck with everything.*

I sat in bed that night, wide awake. I knew NBC, ABC, and CNN were knocking on Castro's daughter's door, but I was the one the family contacted. And that must have been because I'd passed my number to Angie through Facebook. I decided to reach out again.

Hey—how are you doing? I typed to Angie through Facebook Messenger. I held my breath, waiting to see if she'd write back.

She responded moments later. She must have received word that I'd been of assistance on the mug shots.

Overwhelmed, she replied.

I can only imagine, I typed.

While the news stations left gift baskets outside her door, hoping to get her to speak on-air about her father, we established a modicum of trust, and continued messaging into the night. Angie was in shock. As the media waited in a line of trucks, we discussed our families. Sitting on my bed, typing away, I told her about my parents and their messy divorce. She told me about her father, who had, she just learned with the rest of the country, kidnapped these women. I rested my head on my pillow as we connected in a strange way, and she asked for my advice.

What do you think I should do? she typed.

I told her to be careful, to talk to someone she trusted. It took me a second to process my own thoughts, but then I realized: *That person should be me.*

Hesitantly, I made my pitch.

If I came out there, would you be willing to speak with me? I asked the question, having no idea whether CNN would send me. After all, Anderson Cooper was our main talent on the ground.

I'd really like that, Angie responded.

Immediately, I called Erica.

"Let's start putting together a plan. I bet we can convince them to send you," she said, reading my mind. Both of us were concerned I'd be viewed as too much of a rookie to be sent out for such a high-profile story so early on in my career.

Erica and I had successfully extended the tech beat from covering up-and-coming tech startups and their funding to using technology as an investigative tool. A few of our breaking news stories had made it to the weekend news or an afternoon time slot, but I'd never been sent out in the field—especially on a national news story. Those were reserved for seasoned reporters, like the ones I'd once teleprompted for at the Desk.

The next morning, Jeff stopped by Erica's desk and asked her if I had anything new.

"Well, I think Laurie has an exclusive with Castro's daughter," she replied.

Within moments, Jeff called me in.

"What are you waiting for?" he asked, his blue eyes twinkling. "Get on a plane!" As I rushed out the door, I heard him yell, "And change your shirt!"

I went home and packed the black blazer.

As the day turned into a tornado of logistics, there was talk of sending me to Cleveland with a more seasoned producer instead of Erica; someone who'd produced high-profile live shots in the field before. After all, Matt Lauer (pre–sexual harassment allegations) likely wanted this interview, and yet it was going to me, Laurie Segall, who barely had a footprint on-air. I didn't care. I insisted the producer had to be Erica.

"It's got to be her," I said to Caleb, who was the only executive at CNNMoney with power in the TV landscape, with whom I felt comfortable voicing my opinion. "We're a package deal."

Within hours, Erica and I were on the flight.

By the time we arrived in Angie's quiet suburban neighborhood in Cleveland, cars had lined the streets and reporters had crowded her front door. I pushed my way to the front with Erica and knocked.

A young woman with long dark hair wearing a teal shirt opened the door and let us in. We'd developed a rapport on Facebook, so as odd as the circumstances were, she didn't feel like a complete stranger. I hoped she felt the same about me.

I looked around the home's modest interior, which was full of wooden beams and had a warm feeling, despite the dark cloud that loomed over the day.

"I brought some pictures, if you'd like to take a look," she said softly, her Midwestern accent lilting her words.

"Sure," I said.

As our camera crew set up, she showed me snapshots of her and her father. Tears filled her eyes, and I knew I'd have to press her carefully. What did she know? Were there signs that her father was a monster; any clues that he'd kept three women locked away for a decade? I sensed that she was terrified. Her life had quickly become the biggest story in national news.

We settled into two seats in her backyard and began the interview.

"It's like a horror movie," she said to me as we started filming. "It's like watching a bad movie."

"Only you're in it," I said, listening to the birds in her backyard.

A soundtrack of chirping—a sign of joy—against the backdrop of this interview felt almost cruel. I watched as Angie grappled with the clues: often she hadn't been allowed through the front of the house; the basement door always remained locked; and at times, her father disappeared from dinner for long periods of time.

As we sat across from each other, she spoke quietly but with assurance, vowing never to see the man she used to call Daddy again.

"He is dead to me," she said. Her voice cracked, with both exhaustion and determination.

"Your family is attached to this stigma," I said, as the interview was ending. "What is the message you want to tell people?"

"My father's actions are not a reflection of everyone in the family," she said.

Faintly, I could hear knocking on her front door. We both ignored it.

She continued, "We don't have monster in our blood."

We sat quietly for a moment. For a decade, her father had held captive, raped, and beaten three young women in the same home where she'd shared dinners and music with him. There was no going back.

Erica and I would leave her home. The media outside would

eventually dissipate. She would live with the reality of the story every day for the rest of her life. It was heartbreaking and cruel. Journalists got to step into people's lives and their pain, and then step out, leaving it behind. I promised myself that I wouldn't forget her words, her message to the world: *We don't have monster in our blood.*

I didn't want to leave her story behind once I'd left her home. I hadn't worked for many years in the newsroom, but I'd met reporters and producers who were jaded, who walked away from stories untouched by the intensity of the orbits they'd entered. A part of me understood why. It was self-preservation. It was a method of coping. It was how you learned to keep moving forward as you witnessed real pain and horror, but I didn't want to be numb. I always wanted to remember the flicker of pain in Angie's eyes when she spoke to me about the worst day of her life.

As we said our goodbyes, I felt terrible, racing out of her home as her words hung in the air. I didn't want to leave her there alone. But I knew we had to pack up to get the story live on TV in the next hours. My job was to tell her story.

Erica was already on the phone with Anderson Cooper's team, talking logistics, as we walked toward the car. We drove as fast as possible to the CNN Live truck that was stationed outside Ariel Castro's house, fifteen minutes away. We practically fell into the van, where we transferred our video to a bay in Atlanta. There, someone cut the tape, and then sent it into Anderson Cooper's feed in time for him to say, "Tonight, the exclusive interview with one of Ariel Castro's daughters. Take a look at what she told CNN's Laurie Segall about her father's behavior."

A clip played as I positioned the IFB in my ear. My palms were cold with sweat as I prepared for my first "live in the field" interview. What if my voice shook; what if my expressions and body language didn't look right? What if the world could sense that this was my first time under the intense circumstances of breaking news?

I heard Eli from Anderson Cooper's control room check me in.

He'd known me from my teleprompting days as a news assistant. He was the director, and top of the food chain in the control room.

"You got me, Segall?"

"I got you," I said as I stood next to Anderson, who had been sent to Cleveland for the story. I held my reporter's notepad close to my chest, willing my clammy hands not to shake. Erica stood inches away, just out of frame, both of us silently praying I'd do well.

"Laurie joins me now with this interview with Angie Gregg," Anderson said, looking into the camera.

"I can't imagine what this is like for this young woman," he said, looking over at me. "All of a sudden, she realizes, if the accusations are true, what her father has done."

"Anderson, she said it . . . This was a nightmare." *Here we go.*

As I stood looking into the camera, the clip that Erica had raced to edit aired. I heard bits of my conversation with Angie, just hours before, as she described the locked doors of her father's basement, the long periods of time when he'd disappear with no explanation. As we sat across from each other in her backyard, America watched Angie piece together where her father had gone.

"Everything is making sense now," she said.

I stood silently next to Anderson, taking slow breaths, waiting for the clip to finish.

"It's all adding up. And I'm just—I'm disgusted. I'm horrified," Angie concluded.

When it was over, Anderson looked at me. "Fascinating interview. Laurie, thank you. Appreciate that." He turned again toward the camera, before introducing his next segment.

I collapsed into the back of a van, inspired but exhausted. My eye caught the rearview mirror, and I ignored the stress of the past month etched in my pale reflection. This was my breakthrough moment at the network, I told myself. I'd officially crossed over from news assistant to on-camera talent, to someone to be taken seriously. I sat curled in the back of the van, barely able to keep my eyes open, and thought about Angie. I messaged her. She wrote back that she'd watched the segment, and I wondered how she felt seeing herself speaking about her own nightmare.

My phone buzzed. It was Daniel.

"You're famous!" he joked, adding, "I'm so proud of you."

I was grateful for supportive friends against the backdrop of the intensity that came along with the late hours and telling these types of stories. I closed my eyes as the crew outside packed up. How was it possible that I felt both proud of myself and still like an impostor just trying to break into the big leagues?

My interview was picked up and aired on ABC and NBC. People across CNN were scratching their heads, saying, "Wait, isn't that the girl from Business Updates?"

When I got back to New York, I was called in by the CNN Talent Department. Sitting in a small, overly air-conditioned office, they asked what I saw for myself. *This is it,* I thought, stuffing my insecurities into a locked drawer. Erica had teed me up, and I'd hit a grand slam.

CHAPTER 8

Values and Valuations

After our conversation, the talent department had offered me a "talent coach," an official investment in my future. My coach was a large man named Lenny, who was determined to make my tracking voice sound more professional. Every week, Lenny and I spent an hour in the tracking booth to hone my narration skills. And after our trip to Cleveland, it was clear Erica and I had found our reporting rhythm. We were inseparable, every day taking on a new story.

By the summer of 2013, I was a known quantity in the tech world and in media. At a swanky event at the Boom Boom Room on the eighteenth floor of the Standard Hotel in the Meatpacking District, I drank champagne with entrepreneurs under twinkling chandeliers overlooking floor-to-ceiling views of the Hudson River and toasted the success of tech's boom. I felt like I was right where I wanted to be.

That June, Edward Snowden leaked NSA documents, and for the first time, users started to understand that we weren't the only ones riding the tech wave—the government was using tech companies to track us, as well. Erica and I spent days trying to get an interview with anyone in Snowden's circle—after all, it wasn't every day that you risked it all to reveal secrets of the government.

We ended up Facebook-friending his friends in Hawaii, a crew of Burning Man types whose images revealed sand-themed acrobatics and aerial dancing. As I clicked through pictures of Snowden's inner circle, hoping to gain an insight, I was lost in a sea of sunsets and long, flower-filled hair. One of the women, a flamethrower in his orbit, accepted my friend request but unfortunately not my interview request.

It was the beginning of the growing tension between tech and the government, and the seed of user distrust had been planted. But to me, it pointed to something else, too. Technology was becoming the backbone of society—and when Snowden leaked the documents, we began to wrap our heads around global surveillance and the increasingly fine line between national security and privacy. In revealing those documents, he'd introduced a debate that would only mushroom.

Erica and I were now looking at the tech beat from every angle. We'd moved beyond just covering startups and breaking news. Increasingly, we were looking at how technology was seeping into culture. The tips were coming from everywhere: small talk at cocktail parties where an entrepreneur said something in passing; a relationship I'd built with someone over the years who texted me, saying, *Laurie, you've got to look into this.* Sometimes Erica and I would see an article and think, *There's got to be more to that.* We'd start digging, and one call would lead to the next. Eventually we'd have a completely different story than the one we'd initially envisioned.

"Do you think people will think we're looking at porn?" Erica inquired mid-workday as we browsed MyRedBook, a site used to advertise sex.

I looked at the image in front of me: a woman with her legs spread, eyeing the lens seductively.

"Legal is going to have a field day," I replied, shielding my screen full of mostly nude women as Caleb walked by.

I imagined the older men on the Row combing through imagery for our next investigation: sex work in Silicon Valley. Apparently, tech wasn't all about the front men; the new influx of cash meant that there was quite a bit of "action" happening behind the scenes. I'd heard rumblings that high-end escorts scouted prospects at a five-star hotel in Menlo Park frequented by investors flush with cash. It was a starting place. For weeks, Erica and I combed through MyRedBook, surfed Craigslist, and found a trove of material. We began speaking to a group of sex workers who considered themselves Silicon Valley's "other entrepreneurs"—workers in the world's oldest profession taking advantage of the new money pouring into the world's newest profession.

"I think I found someone!" I stared at a Craigslist listing advertising for Bay Area tech geeks interested in being "dominated." "I'm going to respond to her ad," I said, looking over at Erica, who remained unfazed.

Weeks later, I was in Oakland, California, to meet a dominatrix named Madame Rose. Before I took Erica or a camera inside, I'd decided to scope out the place.

When the cab dropped me off, I couldn't imagine that the tidy little building in front of me also housed a dominatrix dungeon. There was nothing notable about the beige lobby, other than how quiet it was. I scanned the nondescript space for any traces of BDSM, but found only signs for the League of Women Voters. I went up the stairs and through a side door that led to a secret dungeon nobody else in the building knew about.

A woman with long dark hair and jade eyes wearing an ankle-length black dress greeted me.

"Welcome," she said, smiling. She led me into the dungeon, full of contraptions, hanging sex swings, and a large cage in the corner. The wall held black leather masks, gas masks, long silver chains, and leather restraints, and what she would later inform me was a device used for low-grade electric shocks. I willed myself to act natural in an attempt to come off as professional and not prudish.

"Take a seat."

I obeyed, looking around uncomfortably. I found myself propped against what I soon learned was a bench reserved for spanking.

"Everything in here is high-tech," she explained, as I glanced at one of the gas masks that had iPod earbuds attached. "My tech clients helped me create much of this."

She pointed to a large contraption above us.

"That crane can hold up to nine hundred pounds. It was built by one of my engineer clients."

I looked up in awe.

Then she led me over to the large iron cage in the corner.

"Want to get in?" she asked, lifting an eyebrow.

"I'm good," I said quickly, wondering how I'd explain to Erica getting locked in a sex cage.

She gestured toward it. "This jail cell here was constructed in perfect proportion to the jail cells in Alcatraz by an MIT engineer."

"Will he go on camera to talk about it?" I asked.

"Definitely not."

We paused for a moment as I studied the perfectly proportioned cage, gleaming against the red walls. I wondered whether any of the engineers I'd encountered at dozens of tech conferences had been inside it.

Madame Rose read my mind.

"Where do you think all the Apple engineers get their creative inspiration? I lock them up for the weekend."

We laughed.

"How do you feed them?" I asked.

She pointed to a slot with just enough room for a tray.

"Legit," I said, impressed. I couldn't wait to attend Apple's flagship developers' conference in Palo Alto with a whole new perspective.

We sat on large leather couches, surrounded by handcuffs and body restraints, discussing power. At first, I felt uncomfortable, but once we began speaking, I was at ease.

"These guys get off on it, they love being controlled," she told me. "Why shouldn't I be able to make money off the boom? I have a respectful relationship with my clients. We have rules, boundaries. No one gets hurt—without consent, of course." She winked. "Want to see the nipple clamps?"

Increasingly Silicon Valley was the center of power. The deals, the money, the ego, the people here who were "changing the world"—it all happened in her backyard. A crane didn't need to hold nine hundred pounds, but because it was built by one of her engineering clients, it could. And maybe that was the point of this. Excess and possibility, defying norms, power and control—I could feel it all hanging in the air, alongside whips and chains.

I wondered out loud about the connection among power, control, and sex.

"Oh, honey," she said, crossing her legs. "If you only knew."

As I left the building, I couldn't help but think, *I have the coolest job in the world—and maybe I'm insane*. I loved visiting other people's lives, no matter how different they might be from my own. I admired Madame Rose's confidence. There was something appealing about her relationship with power and her unwavering ability to own it. But there was something else, too. As more money poured into the startup scene, the community was beginning to feel different—less like a late-night karaoke party, and more like the frat parties at University of Michigan that I'd avoided. Perhaps a part of me felt a certain delight in imagining the overconfident bros of Silicon Valley getting locked up.

Erica and I continued to build out the story, but we had one small problem: although we'd convinced our bosses to send us to the Bay Area to track down high-tech's sex workers, it turned out that many of them weren't exactly reliable.

Erica called me from her hotel room before we were set to shoot with one of the sex workers.

"Becca canceled."

"What do you mean, she canceled? We're supposed to put her on camera in two hours!"

"I know."

I sighed as a text lit up my phone. I read the message: *Sorry, thought about it more. Don't want to go on camera. Good luck.* At least she was polite.

"Shit. Tiffany also just backed out," I said.

"Okay, so I guess now isn't the best time to tell you that Alison is ghosting me?"

For the next day, Erica and I holed up in our hotel rooms at Union Square in San Francisco, browsing MyRedBook, cold-calling sex workers, and answering Craigslist ads. I didn't mind—I was fascinated by these communities, and curious as to how and why women landed there. I had more of a stomach for this than the market hits I'd been writing about for Business Updates. These corner stories were easier for me to grasp. I'd choose to explain the relationship between a dominatrix, power, and Silicon Valley over junk bonds any day.

Thanks to a stroke of luck, we found new women willing to let us take a peek into the lucrative Silicon Valley sex boom. Within hours, we were meeting with a lady who proudly discussed her work and showed us a little piece of plastic: Square. She used Jack Dorsey's credit card reader to accept mobile payments for her services.

This probably wasn't what Jack had in mind, I thought, recalling our interview at the coffee shop.

"I file it under a different business name," the woman explained, her sparkling blue fingernails holding up Square's reader as she swiped a credit card. "As far as Square knows, it's a consulting business."

We hopped into the crew car and drove ten minutes to meet Kitty Stryker, a woman who was a social media marketer by day, and a sex worker in the evenings. Kitty, who wore pigtails and black-rimmed glasses and had molecules tattooed across her forearms, described the ebb and flow of startups in sex worker lingo.

"A startup will be doing really well, and you'll see a bunch of people from that startup. And then they might falter, and then you'll see a bunch of people from a different startup," she explained, as we sat on her couch in San Francisco's Castro District.

Another madam we found was buying up *Game of Thrones* underwear because geek-themed clothes were "on market." She encouraged "her girls" to wear the attire in online ads, to appeal to nerdy guys flush with startup green.

We asked if we could film her with all her underwear and other memorabilia—which included T-shirts with phrases like "Winter Is Coming" and "Geeks Make Better Lovers"—and she agreed.

Erica negotiated with a local business to rent space for our shoot.

"What are you filming?" they asked.

"The story of a female entrepreneur," Erica replied. That was not a lie.

The man nodded. Erica handed him $300 for allowing us to use the space, and we filmed the segment, with the woman explaining how she marketed to new money. Our photojournalist aimed his camera at a table full of geek-themed lingerie, as I discussed her underground business and the implications of an influx of cash in her area.

"Well, I have to say, this is one of the crazier segments we've had on," Don Lemon joked as it aired on the weekend show weeks later. "But I can't say we didn't learn anything."

In November 2013, the tech world was buzzing. Twitter was going public. Back in Manhattan, I raced to Wall Street to cover the IPO, where a giant blue Twitter flag hung over the New York Stock Exchange. As I pushed through the crowd, notebook in hand, I looked around for fodder to include in my segments, scribbling details about the day. The room was brimming with excitement and energy as traders in blue jackets exchanged jokes about hashtags and creating Twitter accounts.

The media stretched their mics toward Jack Dorsey, Biz Stone, and Ev Williams, while anchors relayed cheesy lines about Twitter chirping its way to the public markets. Years back, few of these reporters would have given a startup the time of day, but everyone was paying attention now. These guys I'd known as creatives and dreamers—the same guys I'd met in small coffee shops and made jokes with about their pot-smoking home life full of wild animals—were about to become millionaires many times over. Twitter was officially big business. One of tech's darlings had officially grown up. Many said it was the biggest IPO of the year. But the same question that had haunted Facebook rose from the grave: How would a startup fare in the public marketplace?

Instead of reveling in the moment, I jumped on the subway and raced back to the studio in time for my live segment with Jake Tapper. I sensed the significance of Twitter's market debut and the founders I knew who were stepping into a new era. They weren't Wall Street guys, but today they had become the new kingmakers. I needed to convey that, with all the details, in a couple of minutes.

I ran-walked to the elevator banks, waving to my favorite security guard, Gary.

"Hey, superstar!" he said, opening the door for me as I rushed to the studio, fast enough to get there before the cameras went live on an empty seat and slow enough to save my breath.

In an instant, Erica was by my side.

"Have everything you need?"

"Yep."

"Dorsey made six hundred million today. Good number to mention."

"Copy that."

"I'm heading to the control room if you need anything."

"Two minutes to get my head together?"

"Try a different industry."

Before I could organize my notes and straighten my mic, Tapper was in my ear introducing me on-air. I took a deep breath.

"It's an extraordinary day for this startup-turned-public-company, Jake," I began, finding my rhythm. I wanted to capture the ethos of the moment—the fact that while many founders rang the opening bell themselves, Jack, Biz, and Ev had opted to let some of their most beloved Twitter users do the honors: a young girl who sold lemonade for charitable causes; a police chief; the actor Patrick Stewart.

After I'd finished the segment, I finally let myself breathe.

"What a day!" I said to Erica, opening the studio door. "That was smooth, right?"

I noticed she couldn't make eye contact. Erica had a notoriously bad poker face.

"We have a small problem," she said carefully. "When you ID'd Patrick Stewart on-air, you said he was the actor from *Star Wars* instead of *Star Trek*."

I considered the slipup. Not the end of the world.

"Do you think anyone will actually notice?" I asked.

She didn't sugarcoat it. "I have been informed by the control room that this is one of the worst mistakes a human being can make."

Sure enough, all hell broke loose online. A *Star Wars* versus *Star Trek* mix-up is a cardinal sin and indeed the worst mistake of all time. My Twitter feed was full of vitriol for days.

As I hid at my desk, swigging coffee and popping Motrin, I watched as another intergalactic battle took place—this time, along the corridors of CNN. For years, I had been talking about how tech was disrupting the world, and now, it was disrupting the newsroom. The network, a traditional media company, was struggling to stay relevant in the world of Facebook, Twitter, and new media. As a result, Susan Grant was dethroned from her role running CNN .com, and the powers that be brought in a team run by two former (white, male) Bloomberg executives to usher CNN into the twenty-first century. They were younger and flashier, and they promised to make CNN "cool."

Day by day, the corner offices had been changing hands, and in August, a new era at CNN had begun. CNN.com and CNNMoney fell under the realm of CNN Digital. Then, in September, *The Next List* was canceled, leaving Deb scrambling for a job. She'd spent the last two years traveling from Beirut to Buenos Aires for long-form, character-driven stories. But now, it was back to the drawing board. The show didn't fit the vision of what Deb and I had dubbed "new CNN," and Deb was in jeopardy of losing her spot at the company. She found one on the new digital unit, but having traveled the globe to interview visionaries changing the world, she felt the content and focus were completely different from the stories she'd been excited to shoot and produce previously.

The new digital team's priorities were quickly reflected in the culture of the newsroom. Fewer videos were character based. Instead, more were filled with text. Increasingly, there was a war to fight for two minutes of airtime. And Erica and I—the old-school tech team—were caught in a strange in-between space.

"Let's hope I survive this one," Ross said to me, mindlessly clicking his mouse, editing a piece, as I sat next to him. Both of us were in shock after one of our editor friends was let go.

"It feels like there's old CNN and new CNN. And old CNN has a scarlet letter," I said.

"You're doing great things. Keep it up," he said, mustering enthusiasm. I could see that he was concerned for himself. For the first time I worried about Ross, my North Star, who had guided me through the trials and tribulations of the newsroom.

When I was summoned by the new head of digital to be a part of its first foray into digital video, I was grateful and relieved to have the opportunity—until one of the several producers he'd brought with him from Bloomberg described the job. He wanted me to be in a hip, edgy video about a high-tech bra that "unhooks for love." *This sounds terrible*, I thought, but I wanted to play ball with the buzzy new team, so I said yes with a smile plastered across my face.

During the shoot, I sat against a white backdrop, slotted in between close-ups of the company's stock footage. As I was prompted to say ridiculous things about how getting turned on unhooked the bra, footage aired of a bone-thin model, hair blowing back, in a plunging glittery black bra that—when paired with a smartphone and apparently a love interest—came undone, leaving viewers with an image of a nearly topless woman. I grew increasingly uncomfortable when a young producer encouraged me to "show personality." I nearly walked off the set. I had spent my career honing in on interesting tech startups and using my beat to investigate breaking news stories, and felt passionate about empowering women, but now I was talking about some lame bra that was probably designed by a horny dude who couldn't get laid. Technically, it was designed by a Japanese lingerie company, but that didn't change the sentiment. The digital team used the same shot of a bra unhooking to reveal women's breasts over my narration. I'd tried to make the piece feel less cheap, expanding the conversation into the categories of "body hacking" and "smart tech," but as I watched the edited work featuring me against a backdrop of breast footage, I was ashamed to have my face on the first digital video in this new era.

The digital team felt the opposite. The video was apparently "innovative" and "beautiful" and shot in a "new style." When it was finished, the digital guys were so excited, they called down Chris Peacock.

Chris came up, watched the playback, and then looked at me, his brow furrowed. "Doesn't really feel like you, does it?"

I nodded, letting the tension between the two worlds sink in. When the video came out, I reluctantly tweeted it, wishing I'd had the confidence to have said no in the first place. But I hadn't. I was protective of what I was building at the network, and wanted to cooperate with the people who would likely have a big hand in shaping my future. I felt alienated by the direction of the piece, but if I spoke up, I knew it would likely deter future opportunities. I was ambitious and didn't want to squander my opportunity. But how could I say yes to a topic that I felt was completely sexist? Ironically, a part of me felt that if I didn't do it, I'd allow myself to be pushed aside. I wanted a seat at the table, and resented that I had to make a compromise in order to earn it.

In retrospect, asking the tech reporter who had developed a serious beat to stare into a camera and speak about breasts was clearly sexist. But earlier in my career, I didn't have the confidence to say no to the boys' club.

I feel dead inside, Deb texted one afternoon. *Can you come up to the eighth floor?* I wasn't sure what was going on, but I sensed she was in need of a pep talk.

As I walked into the interview room, I saw an elaborate setup. Deb had spent hours lighting the room and was hunched over a macro lens. I wondered who she'd be interviewing. These types of setups were reserved for prominent figures. But when I got closer, I saw her subject: an Oral-B toothbrush.

"Um, why are you filming a toothbrush?" I asked.

"Laurie, I don't know if I can do this," she said, looking like she was about to burst into tears.

She'd reached a breaking point. After years of filming around the world and taking on challenging breaking news stories that brought her into the field, she had a new set of directives in her role on the digital team: to help shoot sponsored videos that would make the company money.

One of the new lieutenants who followed the digital team leader from Bloomberg had assigned Deb to a sponsored series called Devices. Essentially this meant shooting commercials for items like Wi-Fi-connected toothbrushes.

I watched as she clicked *record* and panned the camera around the toothbrush.

The story would feature the Oral-B smart toothbrush, and Deb would later film a digital correspondent, whom the leader had also brought with him from Bloomberg, brushing her teeth.

"This is so painful," I said, watching her. "They can't be serious."

"They are."

Our collective digital video experiences were just the beginning of the growing tension between old CNN—producers like the ones I'd worked with at the Desk, who were more traditional news junkies—and "the Bloomberg Mafia," who came in with ideas to make the digital landscape "cool." But so far, those ideas ended up with Deb, an incredibly talented photojournalist, being banished to the eighth floor to shoot toothbrush videos, and me looking into a camera to talk about bras. We were skeptical, to say the least.

A bright spot came months later when a producer told me I should do a story on some guy named Ethan from New Orleans, who had a dinner club startup. Not long after she connected us, Ethan and I met inside the shopping mall of the Time Warner Center.

He sauntered in at 3:00 P.M. wearing a baseball cap and shorts, and I couldn't tell if he was interested in me covering his company, or in me. His body language confused me as he told me about his company, PopDine, which hosted pop-up dinners around the country. And while he spouted company details he also managed to subtly ask whether I was seeing anyone. PopDine brought together a community of diners all over the country and supported up-and-coming chefs, and while it seemed like a good idea, when he invited me to a PopDine dinner that night in Brooklyn, I thanked him, smiled politely, and declined. He seemed nice, but PopDine felt more like an events company than a tech startup, and not a fit for my beat.

Later that evening, I was at the Coliseum, CNN's local watering hole, with a group of colleagues who had gathered to say goodbye to yet another coworker. The goodbye parties were ongoing, and awful. Every couple of weeks, a new invitation to celebrate the long career of an exiting producer or photojournalist landed in my inbox.

"To the old CNN," a producer toasted as we took shots of whiskey. I looked over at the bartender, an elderly Irish man with a kind face who poured extra shots for old times' sake. The tight-knit crew behind the bar held decades of CNN gossip.

"Is anyone hungry?" another producer asked. "I'm starving!"

"I know a guy throwing a dinner party," I proclaimed, tipsy from whiskey and hit with a dose of nostalgia as I stood next to a handful of the producers I'd worked with at the Desk. A small crew of us were hungry and ready for an adventure.

The four of us headed to Brooklyn and crashed Ethan's event—a five-course tasting menu. At the end, I thanked Ethan for being a good sport, and told him to let me know the next time he was in town. He said he'd be coming to New York the following week. He didn't tell me he'd be in town to go on a date with me.

Somehow, Ethan kept "showing up" in Manhattan, and the

more time we spent together, the more obvious it became that we fit into each other's lives. Our conversations were easy and fluid. At West Village restaurants, over sushi hand rolls and libations, we spoke about his love of New Orleans and the company he was trying to build: a community for chefs. I told him about my love of storytelling, and the ups and downs of the changing landscape of the newsroom. A few months into dating, he walked me to Erica and Arel's wedding ceremony near Central Park, and after he gave me a kiss and told me to have fun, I wondered if maybe, just maybe, I'd finally found my Arel.

Ethan and I waited for some time before sleeping together. I wasn't concerned; I was sure our easy conversations would translate to physical touch and chemistry. But I was wrong. Our first time felt rushed and awkward, uncomfortable, like two bodies that couldn't find each other. I was shocked.

I called Deb afterward.

"There's no way this is going to work," I said, whispering from my bathroom.

She asked me if I loved him, and I said yes, I believed I did. In fact, I was crazy about him. He'd become my confidant, and the more we saw each other, the easier it was to tell him everything: my adventures at work, my growing frustrations with CNN.

I felt comfortable with Ethan, and while our chemistry wasn't the kind I associated with Boardwalk, the more time we spent together, the more our physical connection grew. It wasn't perfect, but no relationship was, I told myself. We were beginning to build our lives together, and for the first time, I was truly allowing myself to envision spending my life with someone.

By spring 2014, we were official.

As the cherry trees blossomed and cotton dresses replaced sweaters, Daniel and I toasted my newfound relationship. Ethan was friendly and engaging, always finding a way to connect with my small crew: he spoke about business ventures with Daniel, and still photography with Deb. Both Daniel and Deb enthusiastically approved.

In April, at a conference in Times Square where entrepreneurs buzzed over ideas and met with inner-circle venture capitalists, I ran into Dennis Crowley of Foursquare, who sat quietly at a table. I watched as other entrepreneurs approached him for small talk and acknowledgment. Normally buzzing with energy, the party boy looked exhausted, with dark circles under his eyes.

"You okay?" I asked, sitting down next to him.

When I'd met Dennis in 2010, Foursquare had been the hottest thing on the block and had turned down Yahoo's offer of over $100 million. But now, the hype had died, and the company was entering its rocky adolescence. The press had started questioning the business, and the days of Foursquare's success being splashed across magazine covers seemed to be over. Should Foursquare have sold to Yahoo instead of continuing to ride the wave? Was it arrogance? Luck? Timing? These questions were easier to consider in the rearview mirror.

I watched Dennis as he offered strained smiles to younger entrepreneurs and formed polite sentences out of obligation. The Dennis I'd known years before, full of energy and ambition, had been replaced with a different person. The stress was excruciating, and for a founder, it was a lonely type of pressure.

The mood seemed contagious. The following month, I was on camera with Jack Dorsey for the first time since I'd seen him at Twitter's IPO the year before. He was warm but more muted than he had been in the past, and this time, he was surrounded by handlers. We met in a coffee shop in Midtown—another place where Square was doing business.

"You have a punk rock background," I said to him. "You love people that are different. You had blue hair. But now everyone around you is telling you yes."

"I would not go away from that," he said. "I'm still a punk . . . I'll never stop being a punk."

But he was in designer jeans and a blazer covered his tattoo. It was an interesting contrast. He was still quiet and measured, but technology was quickly becoming the man, and whether or not he wanted to admit it, he now sat in a different seat.

As the coffee machine whirred behind us, Jack defended himself using the history of punk. "The punk scene was the scene where, like, people would get up onstage and they would just play and they were terrible. They were absolutely terrible. Then you saw them next week, and they're a little bit better," he explained. "You saw them in a year, and they were the Ramones, right?"

The same was true for open-source software, he explained. People wrote code, and then wrote it again and again. It started out terrible, but got better with every iteration.

"A year from then, it was Linux. Now it's running the majority of every system out there today, for free," he said. "I think that the biggest thing is to really put yourself out there in a confident way, and learn as quickly as possible. That's what I took from punk, and that's why I'll never stop being a punk."

I couldn't let it end there. I eyed his handler.

"You're a punk, but now you're the tech equivalent of the Ramones, right?"

"I don't know about that," he said modestly.

But we both knew that he was a kingmaker in Silicon Valley, and worth more than a billion dollars. There was no denying that it took a real punk to create two wildly disruptive tech companies—to fight the system, create something out of nothing, and get to where Jack was. But no matter what it took, the closer Jack got to the sun, the harder it was to see the people on Earth that he'd disrupted. If startups were the equivalent of the band before the band got cool, the band was now officially cool, and nothing he said could erase the growing tension between where he came from and where he was going. I remembered him saying that he'd hand out Square credit card readers on the street if he had to and guessed that was a quality that helped him get to where he was today.

"When you're young in this, you have to fight, and that brings out a certain scrappiness."

He looked at me, his eyes blue and intense, nodding in agreement. "It's never getting comfortable. It's making sure that we always are questioning ourselves, and we're always fighting. If we feel like we're not fighting, we're not doing anything interesting."

But when I left that day, I had a feeling Twitter would be fighting different battles in the future. That Silicon Valley was on the cusp of a new era. As the first crop of second-wave tech companies reached their inflection points, it was either sink or swim. These companies weren't startups anymore; they were billion-dollar businesses run by teams—and many of those teams were overwhelmingly white and male. Founders started closing ranks. People who had once been incredibly open and optimistic, people who had once drawn outside the lines, were giving scripted answers. Either the minnows were eaten or they were becoming sharks. Like Jack said, technology was getting better with every iteration—but were *we*? For those of us who used it, the impact remained uncertain.

That question lingered in my mind when, once again, I sat down for an interview with Travis Kalanick. Uber had taken off. The concept of jumping into a stranger's car was no longer foreign; in fact, it had become a cultural norm. Travis was bursting through red tape and waging regulatory battles to achieve his mission of getting Ubers on the road around the world.

Kalanick wore a suit now, his hair salt-and-pepper as we sat against the backdrop of the CNNMoney newsroom. He told the story of "techie kids" trying to make something interesting happen. "We have to tell our story and persuade politicians and city officials about why it's important," he said.

"You're also under fire for aggressive business tactics," I said.

I couldn't ignore that just five months ago, Uber had been accused of ordering and canceling cars from a competing car service in an effort to poach drivers.

"If you call an employee at another company and say, 'Hey, look, would you like to work for me?' that's normal. But if that happens in the car service business, oh my god, the sky is falling," he said, growing defensive.

"Do you think, when it comes to competing services, you guys are sometimes too aggressive with your business tactics?"

"Every once in a while, a team in Shanghai or a team in New York might get really aggressive about recruiting drivers onto the system. But at the end of the day, I think the company has some really strong principles, and we feel good about how we act in the marketplace."

The interview intensified from there. He had come to talk about a partnership with American Express, which we spoke about, but he wasn't prepared to answer hard questions. In the last month, two women had been assaulted during Uber rides. The company had a safety issue, so I asked about it. There was a long pause.

"Laurie," he said slowly, almost like a warning. There was a longer, more awkward pause.

We are not friends. We are not enemies. These are fair questions.

My mind raced as he repeated my name. Then he began to remove his microphone.

"I didn't know it was that kind of interview," he said.

"What kind of interview?" I responded.

"A 'gotcha' interview."

I told him that he had an $18.2 billion valuation. These were important questions.

Travis exchanged words with his PR rep, fuming. Erica was behind the camera, eyes wide, shocked by the emotional immaturity of a man running a company now more valuable than many of the Fortune 500 companies.

Sitting in the middle of the newsroom, Travis's chair across from mine, I wondered whether he would simply walk away. He wouldn't answer the complicated questions that came along with connecting the world; he didn't even have a company line. Perhaps something like "We take women's safety seriously and are working on this"? At what point did a founder believe he was above those types of questions? It was the ultimate hubris. I wasn't there to "get" him, but I also wasn't his mouthpiece. His arrogance was astounding.

He decided to stay, and we finished what had become a combative interview. It would be the last one we did together.

It was a moment in which I could pinpoint a fundamental shift in the community I'd covered. As I walked Travis and his PR person out, the silence crackled with tension. *Changing the world is too simplistic a narrative,* I thought, watching Travis walk past security. You don't get a free pass because you say you've made progress or essentially changed an industry. You've got to constantly check yourself and those around you. Where was the accountability from the people who promised us a better world? And why were they so quick to bristle at fair criticism? As Travis passed by the security desk and Gary ushered him out, it became clear we were entering an uneasy new era.

"All okay, superstar?" Gary asked.

"I'm not sure," I replied honestly.

I continued my frenetic schedule, lining up CEOs, COOs, and anyone at the forefront of change. During those months, Erica and I made our way to Facebook's Menlo Park corporate headquarters for the company's ten-year anniversary. I had secured an interview with Zuckerberg's right-hand man, Chris Cox, head of product. A handsome guy with dusty-blond hair and blue eyes, Chris looked like he belonged in the *Social Network* version of Facebook, rather than the real-life version. The interview was a get: like most industry giants, Facebook kept a tight leash on anyone who was important, which meant they didn't let Chris out very much. It was a shame, because he seemed like a decent human being. Not only that, he didn't speak like a robot, unlike most high-profile tech stars, who were now starting to be walled off by PR handlers.

As Chris and I walked through the maze of cobblestone pathways and brick buildings, it was impossible not to feel the same sense of promise and possibility I did as a freshman in college. Facebook's campus was big but not sprawling. Vending machines spat out phone chargers and battery packs next to signs that read, "Move fast and break things." In the cloud-coated sunlight, everything seemed exciting.

When we neared the building that housed Instagram, I asked Chris what kept him up at night as the company entered its tenth year of existence.

"Staying scrappy," he responded, looking both thoughtful and uneasy.

It was obvious that scrappiness was key to building a company. I'd never met a successful entrepreneur who hadn't rolled up his or her sleeves and waded into the arena with grit and scrappiness. But Chris was touching on something larger. I just couldn't put my finger on it yet.

Meanwhile, Erica and I were taking the scrappy approach to surviving the new world order of digital mandates. We were determined to learn the new tricks of digital-style shooting, and combine it with what we knew: hard-nosed reporting. I didn't want to get pulled into any more bra videos.

In order to survive the revolution, where a key objective was to make online video look better, Erica had tapped into some of the newer digital shooters to demo how to use the new equipment. She decided to take it for a test run. With CNN's blessing, she attended a gaming conference in Washington, D.C., hauling around different cameras and testing out techniques the new executives required for quality digital work.

"So . . . I spent the last weekend in a massive arena watching gamers game," she explained, flush with stories of hotel rooms filled with people who watched other people play video games for sport. "I'll never get that weekend back, but no one can say we can't keep up now."

We looked at the footage. Sure enough, it was beautifully shot. She'd started using new lenses and new angles to fit into the new directive: *Make it look better for digital.*

In October 2014, CNN announced that it was laying off 10 percent of its workforce. Just months before, it had offered a buyout to employees over fifty years old with a clear message: if not enough people take the buyout, there will be firings come fall.

And there were. Erica and I held our breath as the producers we knew throughout our careers were called into HR, handed blue folders, and then escorted out of the building. Many of the producers who'd helped train me at the Desk were on the chopping block, too. They'd taken the buyout or were let go. These were my mentors, the people who had taught me to check facts and pushed me to pick up the phone and call people. No, they weren't all that excited about Twitter and Instagram, but they didn't hide behind screens. They ran toward the story, rather than just tweeting their opinions about it after the fact. They didn't care if footage "looked cool"; they cared more about the content. And as the digital team leader's lieutenants took hold of CNNMoney, there wasn't a place for Caleb anymore. He also left. Erica and I were devastated, saying goodbye to an era of the company that had given us a journalistic foundation that we were proud of.

When the rounds of cuts at CNN finally slowed, Ross's job was safe, and Deb was adjusting to her new position at the digital team.

Erica and I still had our desks. We had survived—with a mandate to grow our tech team.

While we struggled through these shifts in the newsroom, Erica and I put our heads down and looked at what was coming next in tech: Invasions of privacy. Hacking. Security breaches. I'd downloaded Tor, a browser that let me explore the dark web anonymously. I'd been fascinated by the concept of an alternative web, hidden behind layers of encryption, and its implications.

So when hackers from North Korea dumped data that contained then–Sony CEO Michael Lynton's inbox on the dark web, I was one of the only reporters at CNN able to access the information. The dark web was a foreign concept to many in mainstream media, and how to access it was an enigma. With Erica, I met with our colleagues and discussed the ethics of the situation. What should we do? Should we download Lynton's emails or not? Report the story or not?

As evening set in and the newsroom began to clear out, Erica and I sat in a colleague's corner office and shut the door.

"Legal cleared us to do this," Erica said after hours of back-and-forth.

"So we do it," I said.

We huddled over a laptop, and as the files downloaded, the inbox filled with hundreds of emails, many sent to prominent figures. Investors I knew. Well-known founders. There was talk of deals to be done, and fascinating behind-the-scenes fodder. Instead of the drippy, preprogrammed answers we got on camera, the emails revealed a waterfall of everything we *weren't* supposed to see. My eyes grew wider as I stared into another person's brain: Lynton's everyday messages regarding work, personal notes regarding his children. As I scrolled further, I sensed the weight of a new era, and the ethical challenges we would face. It was fascinating, and all wrong. I felt both guilt and intrigue. Someone had wanted us—journalists—

to see these emails. But what did they want, and what was our role in it?

Where would we draw the line? What about this whole ordeal was newsworthy, and was it okay to report any of it? The hack itself was momentous, but if we reported it, were we merely puppets?

The next day, I appeared on-air with details surrounding the hack, but I chose not to report any of the contents.

"This is unprecedented," I said, trying to explain the complexity within a two-minute clip. The feeling of violation, the ease by which someone's private moments had been accessed and shared. The fear that this was just the beginning.

"That went well," Erica said, as I slid off the set.

"I still feel like I'm not capturing why this matters," I said, increasingly frustrated by the impossibility of fitting the growing ethical implications of tech's influence into sound bites.

"Segall, just a thought," Erica said. "Maybe we should bitch less in email, you know, just in case."

She was right. I couldn't count the number of emails Erica and I had shared, airing our frustrations about the new digital teams. How many times had we complained about their text-filled videos, empty of heart and depth?

Oh god. If anyone saw our messages . . .

I vowed to be more discreet, hoping it wasn't too late.

Heading into the New Year's break, the newsroom was empty. Phones still buzzed, but only a skeleton crew dotted the open desks. Erica and I were scheming in the secret room, hidden behind one of the control rooms. The sun set early over Central Park, and as we looked at the cars spinning around Columbus Circle, Erica pulled out a piece of paper. Together, we wrote down our goals for the new year.

"We've managed to get you on TV more," she said.

I nodded.

"And we're building out our tech team," she added. "I feel good about that."

I bit the tip of my pen. "I think we can get longer, more-in-depth features on-air," I said. "Let's get beyond two minutes."

"What if we went for our own special?" she offered. I loved the idea and wondered if we'd ever be able to do it.

As we wrote out our goals—people we'd like to interview, places we'd like to go—we topped them off with a priority: get our own digital special series. We'd use the new digital shooters, and come up with a series that people could binge-watch online.

"Harder things have been accomplished," Erica joked as we said goodbye in the nearly empty newsroom, ready for 2015.

CHAPTER 9

Sex, Love, Silicon Valley

What do you mean, there are sex parties and everyone is polyamorous?" I asked the VC sitting across from me, who was also a former Twitter lieutenant. We were having dinner at a fancy restaurant in the Ferry Building, in San Francisco.

"Disruption" was the word in Silicon Valley, and over half a dozen oysters, I was learning that the concept was also being applied to the personal lives of the people coding our future. I knew there was an uptick in experimentation with micro-dosing and smart drugs for productivity, and now, it seemed, there was experimentation with relationships, too. A new app called Secret, built by a former Square engineer, allowed anyone to post anonymously on the app and "share secrets" with their friend group and acquaintances (a ticking time bomb in the internet age). It was funded by major investors in Silicon Valley, and while it hadn't quite become mainstream, the app was serving as a distracting source of Silicon Valley gossip. People were posting everything from VC infighting to rumblings of swingers' parties and startup acquisitions that had yet to take place. It was catnip for journalists, and after three postings referencing a particular sex party, I'd decided to do some digging.

Eyes gleaming, the VC broke it down: "From what I understand, people fall into two camps: the swingers and the polyamorous."

"From what you understand . . . or what you know?" I joked.

He laughed. "Listen, I know people in both camps. Everyone is experimenting."

"Would they talk to me?"

"I'd love to see that," he challenged.

He went on to describe the philosophy and the principles. Polyamorous couples could have multiple romantic relationships. Sometimes, people had a primary partner and other arrangements, meaning both partners had relationships that were secondary, but the configurations varied. The swinger community, he went on, was less about relationships and emotions. Swinger couples went to events together, had a good time, and often had sex with different partners for the evening, but that was where things ended.

I picked up another oyster. Ethan and I had been dating long-distance for almost a year. I was happy with him, and relatively secure, and on a cerebral level, I understood that he or I could be attracted to different people at different times, even if we were in a committed relationship. But the idea of having multiple partners—could I relate to it? Sure. Could I do it? Hard no. If Ethan asked me to be polyamorous, I would throw him off a cliff. But I was curious about why other people found it so appealing.

There was something interesting about the "don't play by the rules" entrepreneur mentality when applied to relationships—even if polyamory and swinging weren't exactly groundbreaking. I started reading about various alternative communities, and learned that in the nineteenth century, in upstate New York, a minister named Jon Noyes conducted an experiment offering a way around traditional monogamous relationships, called "complex marriage." Three hundred people lived in a commune, and all of them were considered to be married to each other. There was also plenty of material from the sixties, when non-monogamy was popularized in an era of free love. But I was curious to see Silicon Valley's take on it.

I pitched an idea for a long-form digital series about the "other side" of Silicon Valley to the newly anointed head of digital. After all, it was his team who'd green-lit the video of the bra that "unhooked for love."

"What do you think?" I asked, after he'd studied a paper I'd typed up with descriptions of all the titillating stories Erica and I would cover as part of the special: polyamory, swingers, the rise of smart drugs and LSD . . . We'd titled it *Down in the Valley*.

I saw his brow furrow. It had taken me weeks to get a meeting. *Oh no. He hates it.*

"I like it. But let's call it *Sex, Drugs & Silicon Valley*," he said.

I nodded eagerly.

Erica was pacing by her desk as soon as I walked out.

"We got it!" I whispered.

It would be the first of its kind—a long-form series—in the new digital era of CNN.

The following month, Erica and I began to shoot *Sex, Drugs & Silicon Valley*. Through a source, I learned that a major founder of a software company who'd sold his company for millions was polyamorous. I asked if he would meet me to chat about a story, and prepared for a graceful pivot to "Are you polyamorous?" We met at a popular spot in the Flatiron District. Walking into the noisy bar, I cursed myself for choosing a place with so many people around. This wasn't the best location to ask a newly minted millionaire whether he had multiple partners, especially when he thought we were there to discuss software.

After congratulating him on his company's sale, I steered the conversation toward relationships, trying to strike the delicate balance between professional and prying.

"So, this is a bit awkward," I began, "but I'm doing a special on different types of relationships in Silicon Valley. You know, people who are opening up more and living by a different set of rules." Barely able to make eye contact, I took a sip of my gin and tonic and willed myself to keep speaking. "Do you have any knowledge of people experimenting with open relationships? Polyamory, perhaps?"

His body language shifted perceptibly. We both knew what I knew. I waited for him to walk out, but instead we had an in-depth conversation about his polyamorous relationships. And then, a surprise victory: he offered to introduce me to his (second) girlfriend, who turned out to be lovely and charming, and ended up introducing us to *her* girlfriend, who appeared in our series.

A month later, Erica and I traveled to San Francisco to begin shooting. On a beautiful day in Dolores Park, we set up our cameras for the interview. I sat across from a woman named Sydney, an engineer at a major tech company who had short blond hair and a calm demeanor as she described her relationship status(es). Sydney was in four relationships. She'd been seeing a woman for two years. "We say 'I love you' to each other," she said, smiling as we sat on the grass. She also saw a man once a month, and kept a slot open for what she called her "distraction spot" for anyone catching her attention. Her primary relationship was with her fiancé, who was a man. It sounded both intriguing and exhausting.

Sydney described how love could be "hacked" the same way traditional industries were upended by people who thought outside the lines. If entrepreneurs could hack transportation, e.g., Uber, why not hack the concept of traditional relationships?

"Polyamory is a form of optimization in the sense that you make trade-offs, and you take risks," she explained. She was a woman speaking openly about alternative relationships, and an engineer speaking in tech lingo.

"In technology, people have higher appetites for risk. Opening up your relationship is really risky in a similar way that starting a company is really risky."

A part of me understood the logic. But opening up your relationship was very different from building a startup to revolutionize laundry. Again, I thought of Ethan. I couldn't stomach the idea. Perhaps I was a uniquely jealous creature, but I didn't think I could ever be okay with sharing in the relationship department.

"Love is irrational. Love is crazy. We do things we would never do when it comes to love," I said to Sydney. "How can you be so analytical about it?"

"I feel completely, irrationally in love with my fiancé." She smiled. "I also have that irrational love with other people."

It was the golden hour in San Francisco, where afternoon sunlight framed the white buildings in the distance. Couples laid out woven blankets and shared bottles of wine. For a brief moment, I could feel myself nodding with Sydney, wondering what free-spirited part of me agreed with her.

Later that afternoon, I strolled down Pier 14 with Chris Messina, a blond engineer with dark-rimmed glasses and a thoughtful gaze. In a sense, Chris had invented part of internet culture, as he was the engineer responsible for Twitter's use of the hashtag. He also described his own relationship status as polyamorous. As we walked along the water, he echoed Sydney's statement, describing his own non-monogamous relationship and emphasizing that traditional relationships were ripe for disruption.

"If you're trying to build a product—let's say, to draw an analogy—and it's failing fifty percent of the time, you might want to consider the design and think about ways of improving that," he said.

Monogamy or marriage wasn't inherently flawed, he added, "but it may not be the right product for everybody." Chris went on to explain that online communities were growing in which people were talking about opening up their relationships.

"So it's like, wow, my weird is not so weird. I could find a community of a hundred thousand people online now, where a couple of years ago, I would have felt like the only person doing this thing."

If the internet was enabling everyone's weird to be a little less so, it was also amplifying everyone's weird, opening up a new era of experimentation. As Chris and I walked by the Ferry Building, I thought about dating apps beginning to take off and the fundamental behavior changes that were happening as a result. Tinder had launched two years before, and already had become a game changer in the dating world; Bumble had followed suit a year ago. Before I'd started dating Ethan, I'd found myself mindlessly swiping through options, and increasingly, at tech parties, people preferred to huddle in corners, shopping for dates online rather than put themselves out there in person. Options that had once seemed finite were now infinite.

"Think of this as dating in the age of 'Big Dating,'" Chris explained, in a play on "Big Data." He explained that the concept of monogamy was established out of scarcity, when resources and mates were more limited. But now we were overwhelmed by abundance. There were options within every swipe of our fingertips, making the concept of "one" less compelling, and "many" more appealing.

I didn't know what to make of his theory. Analytically, I could wrap my head around it, but a part of me wondered about a data-driven approach to something as personal and unstable as our own hearts.

After Erica and I finished our shoots with Sydney and Chris, we flew back home, away from the mecca of analytics and optimization, landing back in the world I was familiar with: New York City and its healthy dose of skepticism. We rounded out the conversation with a cultural anthropologist who was wary of our findings. She told me, "We are naturally jealous animals." Surrounded by books and awards, she defended monogamy from her Upper West Side apartment.

With one episode in the can, Erica and I turned our attention to the swinger community, which many people asked me to refer to as the "lifestyle community."

As I built up my contacts, I learned about a monthly party that a bunch of tech employees attended to "swing." It was hosted by a guy named Ralph, a former tech entrepreneur who, more than a decade ago, had sold his first company for $5 million. If I could earn his trust, I had a better shot at pulling back the curtain on the Valley. I gave Ralph a call, and he told me that my best way into this world was to show up.

"What do you mean, show up?" I asked.

"Come to a party," he said matter-of-factly. "They're very tasteful. I think you'll enjoy it."

As I considered, I thought back to the sex dungeon I'd visited in Oakland a few years prior. Booking an interview at a swingers' party wouldn't be the weirdest thing I'd ever done.

"And bring your boyfriend!" Ralph added enthusiastically.

Over my dead body, I thought, and managed to squeak out a polite "Ha."

Weeks later, Ethan and I were in an Airbnb in San Francisco. While he was on the West Coast for work, he was amused that I'd be attending the party, and took a supportive-boyfriend role, asking few questions. Instead, he invited over some friends to keep him company while I attended what was described as Silicon Valley's premier sex party. As he settled in with his buddies, I found myself considering a very basic question: What does one wear to a swingers' party when not . . . swinging? I chose a black dress in a classic silhouette, not too short, not too long. It didn't really say anything, which was exactly the point.

Arriving at a nondescript building, I was greeted by Ralph. While I'd envisioned his physical appearance would match a suave organizer of what he'd dubbed one of Silicon Valley's most elite sex parties, Ralph was a large man with beads of sweat dripping down his bald head. I'd learned that four thousand people around Silicon Valley were on his mailing list. Many were startup employees, software engineers, and venture capitalists.

Imagine you're walking into any party—just one with fewer clothes, I told myself as Ralph opened the door. But inside, what I'd hoped would be a bit *Eyes Wide Shut* was more "Shut Your Eyes." There was lots of gold drapery and eye-shifting, like junior prom but with more breasts.

Ralph pointed to an iPad where couples were checking in.

"The guy who created our check-in software basically built Oracle!" he shouted over thumping music, explaining to me that everyone was able to connect on the app ahead of time. He held up his iPhone. "We're going to get a UI overhaul soon, too!" He explained that one of the party regulars was an iPhone developer who'd offered to help improve the user interface to create a better "app experience."

As the music blared, people stood awkwardly around a dance floor illuminated by a green light. I had a polite conversation with a VC who wore khakis and a collared shirt. His girlfriend, who worked at Google, was in a sparkling crop top. Everything seemed kosher(ish).

My instinct was to talk tech. I willed myself forward, asking scantily clad couples to share their stories, immediately identifying myself as "Laurie from CNN" so as not to give the wrong impression.

"I remember when we slept with another couple and high-fived after!" crowed a woman dressed in a Britney Spears–esque schoolgirl outfit. Her husband, a former Square employee, blushed. I smiled politely, willing myself not to visualize the scenario.

An hour later, feeling more at ease, I met Greg, a dark-haired engineer with prominent features who described his day job as one involving supercomputers, and Stella, an IT specialist, who had long curly hair, an easygoing demeanor, and a nice smile. The couple said they'd be happy to chat on camera about their open, swinger relationships. Mission accomplished, I made my way to the exit.

But as I headed for the door, the vibe started to shift. Clothing began to fall to the floor. The nice woman from Google had now taken off her top, revealing fabulous breasts. I did everything in my power to act nonchalant. She stopped me to ask if I was enjoying the evening.

"It's been really interesting," I said, trying not to divert my gaze below her shoulders. *What a strange thing, to have a polite conversation, topless.* I managed to excuse myself, and again headed toward the exit, but this time Ralph intercepted me.

"You have to check out the Magic Carpet Fuck Space!" he shouted over the music.

Before I could protest, he was dragging me by the arm, up the stairs, to a scene I will never be able to unsee. A bin of towels sat next to a door that led to a room carpeted in mattresses draped in red sheets and framed with blue pillows. That's where I saw him—well, the back of him. The same VC I had spoken to earlier was on his knees, naked, a giant multicolored tattoo on his lower back, thrusting back and forth.

"Well, that's something," I said to Ralph, struggling for words as my body began to enter fight-or-flight mode. "I'll be in touch!" I fled down the stairs, making a beeline for the door.

When I got home, Ethan was in bed.

"How was it?" he mumbled.

I sat on the corner of the mattress, unable to erase the tattoo now etched into my brain. Was I a prude? As I watched his long legs hanging over the bed, his head buried in the pillow, I thought about the cheap mattresses and hungry looks, the Lysol on the staircase. The commoditization of sex made me feel like a virgin again.

"I just need a second," I managed to say, trying to forget the writhing bodies in the Magic Carpet Fuck Space.

I followed up with Greg and Stella, and we planned to tape at the next event, which was taking place at a private home in Menlo Park. This time, I had backup: Erica. We set up our camera equipment an hour before go-time, and then we waited.

Porn blasted from TVs mounted in rooms throughout the home as Stella and Greg arrived, just as pleasant as when I'd first met them. Stella, wearing a bright blue dress with a plunging neckline, and Greg, in a shiny faux-leather jacket, sat across from me, explaining what they called the lifestyle community.

"Whereas you might go to a bar to pick up someone if you're single, here it wouldn't be unheard-of to do that, even if you're married," he explained.

They described their relationship, pushing back when I described their setup as nontraditional.

"There's a lot you wouldn't characterize as anything other than normal," Greg said. Both agreed they didn't feel jealousy watching the other sleep with various partners. Rather, they enjoyed seeing each other have a good time.

Afterward, I sat down with Ralph, who was a fixture at the parties he threw. He said that business was booming and explained the way in which mobile devices enabled a modern-day "key party" for the community. In the seventies, a "key party" referred to a swingers' party that couples attended where men would drop their keys in a bowl and women would select a key and go home with the man who owned it. The process was entirely random. But Ralph explained that the tech allowed individuals looking to switch partners the ability to choose multiple people they were interested in ahead of time, making the evening less random. As we finished shooting, the doorbell started ringing; the guests had arrived for the evening's activities. It was our cue to pack up and leave.

Erica and I agreed we had the material for an incredible digital series, but we wanted more. As we drove away from the swingers' party, we started stewing on a larger question.

"Do you think we could turn a digital series into a thirty-minute special for TV?" I asked.

Unlike a normal television special, shot by photojournalists for the network, we could piece together the digital bits into a large block of television. No one in the digital realm had done it, much less asked for it. But since we were creating the first digital series, why not ask for it to be made into an on-air special? It was all new territory.

"We've certainly got enough material," Erica said, still visibly shaken by the swingers' party. "We've just got to figure out a game plan."

"Let's scheme."

Ethan remained in San Francisco for a couple of meetings, and after a brief goodbye, I boarded the plane with Erica at SFO's now-familiar terminal. I got into my seat for the long flight back East, placed my laptop on the tray table, and prepared to write out notes from the shoot. As I typed my thoughts on relationships and all the complications that came along with finding someone (or multiple someones), I thought of Ethan.

He and I didn't have the electricity I'd hoped existed, or the emotionally sophisticated relationship I thought I'd end up with. But he was a solid force in my life, a steady, calming figure in my breaking-news tornado. We quickly moved through the weekends and months, and I could see us moving through the years together.

As the plane took off, I thought about the end of my twenties, and the rounding of a new decade.

It had been over a year since we'd met, and that fact alone seemed to mean we were getting serious. He still lived in New Orleans, but he flew in on the weekends, and his stuff was slowly accumulating in my apartment. His T-shirts were folded into tiny squares and tucked into my dresser drawers. Sometimes I looked at those tiny squares piling up and felt both happy and paralyzed. I was happy that as I neared thirty—an age of record, one where you're forced to take stock of it all—I had someone wonderful in my life. But I was gripped by fear that I had become stuck or complacent, and was terrified of the thought of being with one person when the question still remained: Was this Boardwalk?

Daniel hadn't found his soul mate, and Deb and her girlfriend Sarah had broken up. They'd refused to settle for anything less than Boardwalk, but even still, my Facebook feed was filled with engagements, and my deep insecurity lingered. Maybe I'd never have spectacular. Maybe I wasn't capable of it. I adored Ethan; but then again, I had once adored Mike. Why did I have these hesitations I couldn't quite place?

On the plane, high above it all, everything seemed so clear—as simple as a Facebook status: People were happy or not. Photos were filtered to make us nostalgic for carefree love, glamour, and a simpler version of life, without the gray area of uncertainty. But no filter could capture my complicated feelings for my own relationship status. I was confused and uncertain, within the boundaries of "almost . . . but not quite."

When I landed, I texted Ethan. *Landed . . . I love you.* I did love him, I told myself, shaking off my self-doubt.

The next day, back at the office, Erica went to our head of programming, and I set my shoulders and marched into Jeff's office.

"We have amazing footage," I gushed to the president of CNN.

"It doesn't have to just live on digital. Give us the opportunity to create a thirty-minute on-air special out of this."

He looked up at me, amused.

"People can binge-watch online, and then watch on TV—we can cross-promote," I added, trying not to seem desperate.

At the same time, Erica was giving the same pitch.

"Laurie's speaking to Jeff!" she informed the head of programming.

"Erica's speaking to Michael!" I informed Jeff.

Perhaps if they both thought the other approved, we'd have a better shot.

Jeff paused, looking from me to the multiple screens behind me. I could feel his brain computing the risk. I forced myself not to look down. *Be confident.*

"Okay," Jeff said, unfolding his arms. "Let's try it."

Barely able to contain my excitement, I scooted Erica into our secret room overlooking the cars revolving around Columbus Circle.

"Did we just get you on TV for thirty minutes?" she asked with a wide smile.

"We did it!" I collapsed into a chair for a minute, before rushing one floor down to the Desk so I could inform Ross. After celebrating with him, I texted Deb.

Walk? I typed. I wanted to share the news in person. We met in the lobby, exited the building, and headed toward Central Park.

After I'd relayed my news, she smiled.

"Remember Biz Stone and the interview on the red bench?"

"How could I forget?" If it weren't for Deb's help shooting and finding work-arounds, I would have never been able to prove myself early on.

"You've come so far," she said, "Also, I can't believe you're getting CNN to air a swingers' party."

I nodded. "Honestly, me either."

We had a time slot on Saturday evening for a four-part series on swinging, polyamory, micro-dosing, and smart drugs. It would be on TV—not just on digital.

This was my first big original show. It was also the first multi-part digital special at the network, and the first digital streaming special to be turned into a TV special. We hadn't seen it done anywhere else. If it worked, it could be the first of many. Through grit and luck, Erica and I had been able to game the system, getting long-form online journalism to appear as a longer TV special. Until now, there had been a significant gap between a produced online series and one on television. We were bringing those two worlds together.

Erica and I organized a viewing party for the evening of February 8, 2015, when *Sex, Drugs & Silicon Valley* was set to air, barring any breaking news.

I spent the day promoting it, but minutes after I finished my last segment, I received an alert. Caitlyn Jenner (before she had transitioned) was involved in a car accident where there was a fatality.

I think the accident is newsworthy, Erica messaged.

There's no way we go to air, I wrote back from the set, hoping everyone was okay, and prepared to text a chain of people to let them know Jenner's breaking news had pushed our special. We were still in a commercial break, and the anchor sitting next to me was receiving the breaking-news alerts in her ear. We'd know in seconds if our special would still air. Someone in a control room was going to make a decision on whether Jenner's car accident was newsworthy enough to float our special.

"Coming up next," the anchor said, "another side of Silicon Valley that will shock you . . . you're not going to want to miss this."

We were on.

I raced across the street to the Coliseum, to join Erica and Arel, Caleb, Stacy, and the producers who'd helped us along the way. Ethan, Deb, and Daniel were already there, and I knew my mom and dad were watching from afar. My mom had called our cousins in Nashville to turn on the TV, and my dad had let people know his daughter would be appearing in a special on television. They were both proud and supportive, and neither raised a brow when I'd described the topic.

Four relationships!? my mother texted as Sydney's portion aired. I was still in shock that we'd managed to get such a provocative piece on TV—for a full thirty minutes.

Very nice, my dad messaged, likely unsure what to say about the content he and Harriet were viewing.

Harriet also messaged: *Great stories!* After my father's health scare, we'd started building a relationship, texting sporadically and staying in touch.

Looking around at the crew of people who had become my New York family, friends, and coworkers, and knowing my family was supporting from afar, I felt an overwhelming sense of appreciation for the hard work that had gone into building the offbeat community that surrounded me. They didn't flinch watching out-of-the-box stories air on the network. If anything, they were a part of bringing those stories to life, and everyone in the room had helped me grow. I felt a sense of belonging.

I looked over at Erica. She and Arel were beaming. I smiled at her, and we both thought the same thing: *This is only the beginning.*

The special was widely viewed as a success. Multiple outlets picked up our reporting, ABC aired a piece based on the special, and our stories were topping CNN.com, as more people became interested in Silicon Valley's alternative lifestyle. Inspired by the positive feedback, Erica and I agreed to start thinking about our next project.

"There's something here," I said to Erica, holding up an article about a woman whose daughter had her private photos stolen and sold to a site devoted to what's called "revenge porn." A hacker had randomly breached the young woman's account, stolen topless photos she'd taken of herself—that she'd never planned to release—and sold them to a site on the web devoted to this type of harassment.

"You're right," she said, scanning the piece. "What do we know about this?"

I did a quick Google search. "There's not much on it . . ."

Her brown eyes sparkled. "Let's dig."

In a few days, we had a much better understanding of revenge porn—a horrific form of harassment, mainly directed at women, which was growing in popularity. While many cases of revenge porn began with vindictive ex-boyfriends committed to shaming and humiliating women, there was a growing business in stolen nudes, and in the young woman's case, she didn't know the man who'd hacked her, yet her photos still ended up on one of the most infamous revenge porn sites.

At the time, "revenge porn" wasn't a term in the dictionary, and since law enforcement didn't know how to define it, the government was slow to enact laws, and tech companies weren't moving fast enough to deal with the growing problem. It was a perfect storm of power, sexism, and online harassment, with real-world implications.

We searched for lawyers who were taking on these types of cases. There were few at the time, but the ones we spoke to were helpful, wanting to call attention to the problem. They introduced us to their clients, who agreed to speak with us. And the more women we spoke to who dealt with this type of harassment, the clearer it became that we had our next investigation. One woman, whose ex-boyfriend uploaded naked photos of her onto the web without her consent, described the feeling as similar "to the feeling of getting raped." I nodded as she told me that it felt like a million people were watching the most intimate moment of her life. She lived with constant fear, insecurity, and a lack of control of her image and identity.

I had conversations with more and more women who shared similar stories, and I kept hearing similar themes: Law enforcement wasn't helpful. They said there was nothing they could do. The proper laws didn't exist to protect victims. Women were told they shouldn't have taken the photos of themselves. Instead of receiving help, they were shamed.

Erica and I were determined to call attention to these women who refused to be victimized by their abusers, the algorithms that had trapped them, and the laws that had failed to safeguard them. While we never spoke about it specifically, Erica and I were attracted to stories about women who were harassed, whose situations didn't seem fair. We felt a need to shine a spotlight on the discrepancies in the narrative, because we knew from personal experience that there was a different playbook for women than there was for men.

Both of us were rising through the ranks and learning firsthand the meaning of "death by a thousand cuts." Once digital had changed hands, there were increasingly more men in powerful seats; some of the most outspoken women we knew, like Stacy and Susan—my original advocates—didn't have a place anymore. Stacy had been a valuable part of the newsroom, and then an integral part of our tech coverage, editing with speed, and not only sharpening all the writers, but also coming up with strategy to integrate CNNMoney and the television side. She'd outgrown her initial role, and as she advocated for many women like me, she'd watched as men were promoted ahead of her. She was told to wait, but many years later, she sat in the same spot. She felt she had no choice but to leave if she wanted to rise. And with Susan long gone, the corner offices were filled with the mostly male digital hires from Bloomberg. There was an overwhelming sense of frustration among the women I knew. It wasn't just women who were leaving; the men who'd supported us were also leaving. And as if sexism in the newsroom weren't enough, we covered male-led tech startups funded by male investors. We lived between a boys' club and a frat house. Rumblings of harassment were as quiet and persistent as white noise.

Our experience of sexism was so pervasive, it was impossible to define, and so Erica and I vented our personal frustrations in the only way we knew how: the stories we selected.

We scheduled our first interview outside of San Diego, with a woman named Nikki. We met her one early spring afternoon in 2015, and I was struck by her beautiful dark hair and green eyes. She shared her story for the first time—about her now former boyfriend, a man she sarcastically called "Mr. Wonderful," whom she met when he'd offered to help her with statistics homework in college. One evening, while over at his home, she'd noticed a red light blinking in the darkness of his room. She walked over to explore and discovered it was coming from what looked like a pen on Mr. Wonderful's desk. But when she unscrewed the cap, she discovered a memory card inside full of images of herself in moments she thought were private—from changing her clothes to watching TV. It wasn't a pen; it was a "pen cam" that was recording her most intimate moments. She learned Mr. Wonderful had also hidden other cameras around her home. They immediately broke up, and she began to learn the extent of the damage he'd done. She found out he'd projected her naked images onto various online sites, and attached her personal information. The images were shared across the web without her consent or control. At one point during the interview, she pulled up her computer and revealed a spinning sphere that Mr. Wonderful had created online with her images. We sat on the couch as she read out loud the tags under her images, taken without consent.

"'Dirty girl. Loose whore,'" she recited, almost matter-of-factly, as she walked me through the images.

I minimized my shock and disgust, knowing the cameras were rolling and not wanting to call attention to my horror at a moment that should be hers.

"I would stay up almost all night every night, just in a little cave of a dark hole online, finding more and more and more," she explained. "Law enforcement straight up told me, 'How do you think that we can possibly help you?'"

Nikki said that every time she met a stranger, every time she applied for a job interview, she wondered if the person had seen her naked. She had lost the ability to walk into a room with confidence.

"What's it like to have no control over what someone sees about you?" I asked.

She paused and looked directly at me.

"Have you ever heard of dignity?"

I nodded.

"I lost that, and I didn't think I would ever figure out how to regain . . . respect for myself, a feeling of safety or self-worth. You just want to hide."

I recognized the psychological toll the harassment had taken on her. As we sat on her porch, the sun pouring in, she answered my questions with urgency, describing job interviews where she "wanted to crawl under a rock and die." Instead of listing her accomplishments, she stared at hiring managers and wondered: *Do they know? Should I get in front of it? Should I mention that naked photos of me are out there, that it's not my fault?*

From the safety of her home, sharing her story with me was part of her process to regain the dignity that had been taken away.

Mr. Wonderful had utilized Nikki's image for power and abuse, and neither the law nor the tech companies had protected her. How unfair that he was free to go about his life, while she felt she had to hide because of what he'd done to her.

"If you could look at him right now, what would you say to him?" I asked.

For the first time in the interview, she started choking up.

"'Thank you,' I would tell him."

I didn't know what to say. My instinct was to blurt out, "This man is awful. He's the worst. Don't thank him." I paused, and then forced myself to sit in the uncomfortable silence. I was beginning to learn that as an interviewer, sitting with silence was one of the best ways to let a person open up, to tell you how they *really* felt. *Don't fill the silence,* I told myself, as seconds that felt like hours passed.

Tears were forming. It felt cruel to not reach out to her, but I knew this was her moment, not mine. She hadn't cried when she discussed her naked images blazoned across sites with the words "bitch" and "whore," but now when she spoke about gratitude, the tears came.

"Thank you for absolutely forcing me to become the most amazing version of me that I never would have expected."

I realized I wasn't sitting next to a victim; this was a woman refusing to allow her voice to be taken away. As she spoke into the lens with raw emotion, I could feel her taking back control. She seemed extraordinarily powerful.

"I love me for the first time in my entire existence, and it's because of the character building that I was forced to do because of this." Her green eyes were glistening with tears, but she was smiling.

I stayed silent for a moment and then ended the interview. Again, I felt an overwhelming desire to hug her, but held back. Telling her story was enough.

When I interviewed Nikki's lawyer, Elisa, about the legal hurdles in this case, she told me that states were only now enacting laws against this type of harassment. In the meantime, she explained, the only other way for victims to protect themselves was to sue for copyright infringement.

"Wait a second," I said, processing what she was saying. "Doesn't that mean you'd have to copyright your images?"

"Yep," she responded. "That requires some work."

I could see Erica connecting the dots, too. We were both on the same page.

"Does this mean you'd have to submit your naked photos somewhere to protect yourself?" she asked.

"Yes," Elisa said. "The Copyright Office."

We were stunned. These laws required victims to send their naked photos to the government in order to defend themselves.

Elisa said that if you were a victim of revenge porn, and your naked photos were put out on the web, you could file a take-down notice with individual websites, but that didn't always mean the sites would comply. You could sue, but to do that, you had to register the copyright to each photo, and that meant sending pictures of your naked body to the Copyright Office in Washington.

I let that concept settle in. *If I wanted to protect myself, I would have to send a naked photo of myself to a stranger in an office in D.C.*

Nothing about that felt right. The whole system was disjointed.

"It's an epidemic," she said. Her inbox was filling with requests from people who were dealing with what she referred to as "nonconsensual pornography."

Back in the newsroom, Erica and I couldn't get over the idea that women had to copyright images of their breasts to protect themselves. It was everything that was wrong with tech's growing influence. Social media and the internet were unchecked and unregulated, and they were making this type of harassment possible. The government couldn't keep up; it felt like the Wild West, with antiquated laws unable to help those most in need.

"It's insane," Erica said.

"We should try to show how the system breaks down," I suggested. Maybe if people could see the absurdity of a woman having to defend herself by further exposing herself, it would move the dial. "Think we could do it undercover?"

"I think Washington, D.C., is one-party consent."

I nodded, understanding the implications. One-party consent meant that I could record a phone conversation without the consent of the other person on the line. It was the perfect way to reveal how the laws were light-years behind technology.

One day later, Erica aimed a camera at me as I dialed the number for copyright registration.

As I prepared to ask the Copyright Office if I'd have to send naked photos of myself in order to protect myself, Jeff walked by.

"What's she up to?" he asked Erica, seeing the camera pointed toward my desk.

"You don't want to know," she replied, unsure how to tell our fearless leader that I was inquiring about sharing naked images with the government.

"Fair enough," he said, and walked away.

After a few rings, a woman answered.

"This is the Copyright Office. May I help you?"

My heart beat faster. I didn't want to screw this up; we only had one shot at getting it right.

"Hi! If I wanted to register the copyright of some images of myself, what does that entail?"

The woman told me that whoever took the images owned the rights to them, but I cut to the chase.

"What if they were naked images?"

There was a pause.

"No, the office has no issue with that. Of course, we don't have naked children."

Thanks for clarifying.

"What if the photos were posted online without my permission?" I asked. "Would there be any specific protections for me? Could I redact them?"

"Let me ask, because I'm not sure about that," she said.

I waited several minutes before the woman came back on.

"Thank you for your patience in holding. I called one of the visual arts specialists," she said. "They actually said they do not have a method for redacting photos, but I was telling them your situation."

She lowered her voice. I could sense that she wanted to be helpful and was about to give me some advice. "They said, 'Actually, a bit of advice is, just send the photo as it appears. It would be a stronger case in an infringement.'"

Elisa was correct. There were few safeguards for victims who'd been exposed, and in order to protect themselves, they risked further exposure.

"So it would be in my best interest to upload my naked photo," I said loudly, to clarify, hoping the rest of the newsroom wouldn't hear me bringing home the point.

"Right," the woman said.

"Got it, thanks."

I hung up, knowing we had what we needed: a real-world example of the tension between technology and the law, and what the impact was, especially on women.

Over the following weeks, Erica and I spent mornings and evenings preparing our next thirty-minute special, which was to air on another Saturday evening. Our second special had immediately received a green light from Jeff. It was called *Revenge Porn: The Cyberwar on Women*. As part of the special, we'd feature Nikki's story, my conversation with the Copyright Office, and an interview with a hacker who'd stolen nude photos of a young woman.

The show aired on TV, and the digital pieces topped CNN.com for days. But most important, Nikki's name would soon be attached to a different Google search. It would appear as the story of a woman who refused to accept this type of harassment. She would call attention to a growing crime and bring awareness for all the women who followed.

CHAPTER 10

Bits & Bytes Meet Flesh & Blood

I slammed the door of the taxi, already ten minutes behind to meet my father and brother for dinner in Gramercy Park, when the phone rang. It was an unknown number.

It's a source, I thought, rushing past the gated park and brownstones.

"Hello?"

"Laurie." The voice on the other end surprised me. Whoever was speaking was using some computer modulator to disguise themselves. It sounded like I was speaking to a robot.

"Who is this?" I asked, my heart pounding.

"Who do you think this is?"

I tried to remain calm. I'd been on camera now for years, and every so often, a viewer would email something inappropriate. On occasion, I'd have to block people on Facebook and Twitter. At one point, a lurker had found a personal blog I'd written years ago, when I was in college, which I had posted on Blogger, the site that Ev Williams created before Medium and Twitter. It outlined raw, vulnerable thoughts of the day my father remarried: how I felt like an outsider in my own family when I watched my father marry a woman I barely knew. I described stepping into the bathroom of my aunt Hazel's house and collapsing on the side of the tub, tears streaming. I wrote about my new stepsiblings, who wore crisp khakis, strappy sandals, and warm smiles; I didn't even know their names.

This was before social media transformed the early internet. I had forgotten the blog was there, until someone with a pseudonym and an avatar tweeted me a link to my childhood. It hadn't been malicious; rather it was eye-opening to see such raw and vulnerable thoughts unearthed to me as I'd slowly started building a public presence.

"Okay, this isn't cute," I said, ready to hang up, worried if I stayed on, my voice would start trembling.

"I will smash in your head and kill you."

Then I heard a click.

I walked into the crowded restaurant. New Yorkers dressed in silk shirts with glossy hair laughed loudly over glasses of wine. I saw my dad and brother in the corner of the room, and as I made my way toward them, I heard the digitized voice over and over again: *I will smash in your head and kill you.*

I relayed the conversation to my father and brother as soon as I sat down. Neither knew what to say. I looked over at my father, wanting him to reach out and wrap me in his arms, tell me it was all right. Instead, he sat stiffly, his napkin on his lap, trying to find the words. Our relationship was improving, but he still couldn't offer that kind of comfort.

For days, I looked over my shoulder, locking my apartment door behind me as quickly as I could. I made a list of everyone my reporting could have possibly offended; the list was long. I landed on Defcon—a conference I'd just attended in Vegas, where hackers from around the world had gathered to "show off their party tricks." I'd met some folks who likely didn't play by the rules and enjoyed trolling journalists. The call had completely freaked me out, but nothing more came of it, so I filed it away as a prank from one of the not-so-good hackers.

Many of the hackers I'd met liked to show that there were cracks in the system, that things could be broken, and sometimes they needed to be broken in order to be put back together. They viewed their actions as a service. Our world was becoming increasingly vulnerable, and these people, who introduced themselves with strange code names, had been raising red flags. More often than not, they were met with resistance from companies when they tried to report bugs in their systems, but over the years such companies were beginning to realize that it was important to get these hackers in front of security flaws.

Not everyone hacked for good purposes, of course. There were the gray-hat hackers, those who fell in between, and on the opposite end of the spectrum, those who hacked for malicious or mercenary reasons. As the world became more connected, and companies struggled to secure their sites and devices, there was a growing underground market where people could make a lot of money selling information on the web. Hacking was like a superpower that could be used for good or evil. And I was fascinated by what made a hacker choose one over the other.

During one interview at the conference, I had met a security researcher, a white-hat hacker named Jean, who had thick dark glasses and double-pierced ears with silver hoops. His company found vulnerabilities in code that could leave companies open to attacks and reported them immediately.

"Do you ever think, What if I went the other way?" I pressed him, as we sat across from each other, cameras rolling in the middle of the conference space.

"Yeah," he said honestly. "There's often the fleeting thought I could be on a beach drinking out of a gold-crusted goblet, but I always keep in mind, for any hack . . . there's a victim."

He explained he'd been in situations where in three keystrokes, he could have complete access to a company's database or someone's personal information.

"But there's always a little angel on one shoulder," he said, pointing. It told him to remember what would happen if someone did it to him.

That evening, Erica and I had attended a party filled with hackers, tequila, and debauchery. As we walked in, fire dancers juggled bright flames against a backdrop of palm trees that stretched over slot machines. Before we could introduce ourselves, a large guy stripped off his shirt and jumped into the decorative pool, creating a huge splash that drenched nearby guys talking Python and Ruby on Rails.

"That guy hacks foreign spies," a guy in glasses and a Hawaiian shirt said to me as I stared at the spectacle.

"The guy who just cannonballed into the pool?" I looked at him pulling up his swim trunks, which had fallen dangerously low post-dive.

"Oh yeah, he's insanely talented. He's taken on China, Russia, you name it."

The next day I was staring at a large billboard called the "Wall of Sheep" that displayed the hacked names and emails of all the people who had forgotten to turn off their Wi-Fi—a rookie mistake made by people attending the conference for the first time.

"This is pretty alarming," I said, scanning the names of unsuspecting victims.

"I hope you turned off your Wi-Fi. It's 101," said a guy who refused to tell me his real name but referred to himself as "Squirrel." "Here, it's not what are you drinking; it's what are you hacking?"

As Erica and I had navigated the conference, we continued to attend official events held by everyone from government agencies to Facebook, where VIPs waited to meet Facebook's new head of security, and a tall man named Moxie with blond dreadlocks spoke about the importance of encryption. He would later start an encrypted app called Signal.

We were surrounded by fascinating, brilliant people, but as soon as I introduced myself as "Laurie from CNN," a wall went up. It was clear that the media were not beloved by many in the community. I kept trying, and found that speaking kindly, and expressing genuine curiosity, was my gateway. Soon, one person connected me to the next, and invitations to unofficial conference parties started rolling in.

As Erica and I were ushered deeper into the community, it was impossible to ignore that we were constantly in rooms full of men. There were the outlying women who navigated the drunken late nights and off-color jokes, but they were the exceptions—and notably so. One name that kept coming up was Shama, a security professional who protected the Nasdaq from hackers. We were told we had to meet her.

We tracked her down, and after a brief meeting with the petite blond woman whose electric energy stood out in a sea of lanyard-clad men, Erica and I knew we'd found our next special. We'd call it *The Secret Lives of Superhero Hackers.*

Within days, we made a six-hour drive from the Nevada desert to meet Shama at a skydiving range, where she said that she felt most herself. I watched her as she geared up to jump from the plane, joking with us as she propped a large parachute bag on her slender shoulders and placed goggles over her blond hair.

"You sure you don't want to join?" she asked for the fifth time. Had it not been for my paralyzing fear of heights, I might have.

I watched from below as she opened her parachute and glided toward the ground, smiling at me when she landed, flickers of excitement in her eyes.

"It helps me feel alive," she said, the breeze dancing around us.

It wasn't about the adrenaline, she assured me. "It's about control and freedom." Freedom from a world full of abuse and darkness, beauty and love.

Later that evening, we met at Shama's house for our interview. She poured a shot of whiskey as Erica set up the cameras. "Sorry, I need this." She laughed, throwing back the drink.

Then we began. Shama talked carefully, describing her childhood. Her parents raised her in an ashram in Austin, Texas, which, according to her, was a cover for horrific abuse. The head of the ashram touched her again and again. She was barely eleven years old. Her parents didn't stop the abuse. Instead, she told me, she was told to "enjoy" it.

Shama hid behind the soft glow of a computer, and in the virtual world, she discovered her love of breaking things to put them back together. She learned to hack, to speak a language that opened doors. It took her far from the touch of a man who would later be convicted of twenty counts of pedophilia, mainly due to her courage to stand in front of him and her entire community. The man would later post bail, paying $11 million before he disappeared.

As she grew up, a young woman in a man's world, she carried the weight of the abuse and channeled it into what she loved—computers. There she finally found security, control, and a passion for protecting the vulnerable. She was hired by finance and media companies, and became one of the good guys in the fight against cyber abusers. She could have chosen darkness, but instead she chose the light.

By the time Erica and I landed back in New York City, we had hours of footage we knew would make a great special. But before we could dig in, yet another hack blew everything up.

It was unlike any hack I'd covered. Everything about it felt more intimate. Including the fact that the data was released on my birthday.

After a day organizing my notes from our prior shoots and chasing leads, I managed to shower, get dressed, and find my hairbrush. In the balmy August heat of New York City, Ethan and I walked to one of my favorite restaurants in the West Village. Sant Ambroeus was located on one of the most beautiful tree-lined streets in the Village, a few brownstones away from Carrie's home in *Sex and the City*.

As we settled down, my phone rang. It was my new boss, who'd been brought in from Bloomberg as part of the newsroom reshuffle. He knew it was my birthday and wouldn't have called unless it was important.

"They did it. They published the data," he said. "Can you access it?"

He was referring to a group of hackers who called themselves "the Impact Team." In July 2015, they'd broken into the database of the website Ashley Madison—an online dating service for people who were married or in relationships but who were seeking other, "discreet" relationships. A list of names of people who allegedly accessed the website had just dropped. The hacked data was salacious and personal, offering a window into other people's most intimate relationships.

"I've got to deal with this," I said, as the waiter came up to the table with a bottle of champagne to celebrate another year around the sun.

On a stoop nearby, I searched my phone for one of the security researchers I'd met at Defcon, who could help me sift through the database, which wasn't easily accessible yet. It was currently available on the dark web, but Tor wasn't exactly easy to access either, especially from my iPhone. Once I'd pinged the security researcher, who agreed to help, I texted my new boss, *Working on it*. I returned to the table for an hour to celebrate, even though my mind was somewhere in the dark web.

The following day, I got to work early as the scandal took the newsroom and the nation by storm. It was one of the most compelling hacks in history, touching on human nature, relationships, our secret desires, the lies we tell the ones we love—and the lies we tell ourselves. Colleagues pulled me aside to whisper, "Is so-and-so on there?" Prominent figures were outed as cheaters, and rumors of affairs and marriage dissatisfaction became water cooler talk. When no one was around, I looked for Ethan's name. I knew it wouldn't be there, so why did I look? As I combed through data that represented deceit and dissatisfaction, I wondered if a part of me didn't fully trust him, or myself. But I pushed aside the thought as I pieced together the story.

The hackers claimed that they exposed the database for moral reasons—while the hackers I'd met at the Defcon and Black Hat conferences hacked to expose vulnerabilities in connected devices. Was this moral? Who were these hackers to decide? I believed that cheating was wrong, but my heart broke for these people whose lives were about to be destroyed publicly when their names hit the web, whose families would be painfully exposed.

Weren't we all living a lie in some way or another? How did a data dump turn into judgment day?

As I dug in, it seemed that everyone was, in some way, touched by the scandal. I spoke to a woman whose husband was a pastor who'd committed suicide when his name was discovered on "the list." I made cold calls to those who'd signed up on the website. Over and over again, I heard stories of people who didn't want the life they were living, who felt trapped. There was shame, humiliation, embarrassment—and that didn't even cover the devastation of the partner on the other side of the equation. I also felt for families on the other side of the hacked list, the children and spouses.

Erica and I pulled together the interviews for a condensed TV segment. Racing against the clock, I tried to go beyond the shock value and into the heart of the hack, and what it exposed. Why had this website—which promised those who were unsatisfied in their relationships a discreet way to cheat—become so popular? What was this saying about our society, and how we felt about monogamy and trust? What were *so many* people seeking when they logged on?

I understood what it was like to feel trapped by the expectations that society put on us. I was in the phase of my life where many of the people I knew were getting married or already married. I knew that Ethan and I *should* be thinking about whether or not we'd get married.

I also knew we lacked intimacy. While we often found ourselves sitting down for meals together and sharing our latest work adventures, speaking about trips we wanted to take and things we wanted to do, it felt difficult to go much deeper. I thought about the Ashley Madison hack and the moral dilemma it presented, and my empathy for all concerned. I had a slightly different view of the hack, one I felt was more nuanced, maybe even unpopular. I didn't want to feel compassion for the people on those lists, but the more I spoke to them, the less judgmental I became. I heard stories about hiding and running, about secrets, and about family ruin every time I cold-called a name on "the list." When the hackers released the information, all parties lost. I wondered how Ethan felt about it. It struck me: I wasn't even sure that Ethan would know where I stood on the Ashley Madison situation, or whether I'd know his feelings about the moral dilemma. I felt there was a gray area, that we were oversimplifying this painful hack and its impact. I wondered if he felt the situation was black-and-white, as did so many others, as names were released and families were torn apart. I wondered why I didn't ask him how he felt. I realized we didn't know those deeper layers of each other.

After a year and a half together, long-distance was becoming increasingly difficult. I didn't think I would ever cheat on Ethan, but the idea of going "all in," giving up complete control, was a foreign concept. I wasn't sure I wanted that—or maybe, I just wasn't sure that I wanted that with him. I felt something was off. I couldn't put a finger on it, but something was missing. The same feeling of "almost . . . but not quite" seemed to have resonated with people around the world—the concept mushrooming into a multimillion-dollar business promising a discreet haven for people with a foot out the door or a desire for something more, and prompting hackers to burn down the house, destroying untold numbers of lives.

Battling disillusion and loneliness, Deb and I decided to crash one of Daniel's dates. Crashing one another's dates was an obnoxious friend-group tradition that probably helped explain why so many of us remained single. I hopped the subway to Miss Favela in Williamsburg, an easygoing Brazilian restaurant near the water that turned into a dance party on Sundays. It was where Daniel took Julie, a tall, willowy brunette, on their second date. We sat down with a total disregard for their privacy and quickly felt out of place. For the first time in seven years, it felt like I was intruding. I realized that this woman was different.

Julie was incredibly stylish: she could wear a fedora without irony, and looked like she belonged in a magazine. But it was more than that. In the month that followed, as she joined our dinners and outings, bringing with her a sense of calm, she became Daniel's "nonnegotiable" and "exception to the rule." I watched his face soften when he looked over at her. Months into their courtship, Julie lost her mother to breast cancer. Daniel, whom I'd never seen fully commit, never left her side. He'd lose his mother unexpectedly just four months later, and Julie would be glued to his arm as they navigated another cemetery together. Loss was every part of their emerging relationship, and because of that, they moved quickly, leaning into each other's pain and grief, becoming pillars for each other.

Julie was the perfect counterbalance for my friend. From the moment they met, I saw something in Daniel settle down, as if he'd found home. It was that feeling of connection and authenticity I longed for. It was a feeling social media continued to promise, but I was still waiting on it to deliver.

In November, I headed to Wall Street to watch Jack Dorsey ring the opening bell for Square's IPO; it had been two years since I stood in the same spot when Twitter went public. While everyone clamored over the meaning of mobile payments, I watched Jack. He was polished, in fitted jeans and a suede jacket. Like Silicon Valley, he had gotten a corporate makeover. I thought back to my first interview with him, when I'd sat in a booster chair while he had told me that mobile payments were the future.

He rang the bell, and seconds later, he reached for his mother, who stood next to him, overcome with pride that her son was taking his second company public. I was close enough to see tears in her eyes.

Hundreds of smartphones stretched in the air to capture the moment. People live-streamed and tweeted, posted Vines and Instagrams. The anti-establishment renegade had become the darling of Wall Street. But in this moment, with his mother, he was just a human being.

After he'd shaken hands with the who's who of the New York Stock Exchange, we spoke on camera, this time on the cobbled lanes of Wall Street. Square vendors lined the sidewalks nearby. A huge white flag was draped over the NYSE with the names of vendors using Square: an homage to Jack's vision to help small businesses make money. It read: "The neighborhood is going public."

"Talk to me about what that means to you," I said to Jack.

"We brought our merchants here on Wall Street, and they're selling," he said. "That's how we started six years ago: giving sellers tools to make the sale."

We didn't speak for long, and there wasn't much from that interview that struck me as groundbreaking. Jack wasn't terribly interested in talking about his products and pushing them out there. After all, they'd gone mainstream. Constantly promoting them was no longer a requirement.

That day, I was more impressed by what remained unsaid. Five years earlier, against the backdrop of whirring espresso machines,

Jack had quietly held up a plastic square and promised to do something interesting. That promise had paid out, shaping business—and culture.

By December 2015, the whole world seemed to be living on their phones, logging tweets and likes on social media. Twitter had over 300 million monthly active users on the platform, and Facebook was skyrocketing, with nearly 1.5 billion users.

On a frigid day, I met with Jack Dorsey's former coworker, Ev Williams, in a bookstore in Soho. There was something ironic about interviewing the former Twitter CEO in a bookstore, surrounded by printed words that were quickly becoming artifacts. But, of course, the choice of location was every bit as intentional as Ev's latest venture. Medium was a communication platform he'd created that tapped into a growing desire for long-form content, for analysis, for a safe haven from the noise blasting from tweets, streams, likes, and timelines.

A direct reaction to the real-time nature of Twitter, Medium had hit a nerve, and now, around three years later, already had 25 million to 30 million monthly visitors. Bill Gates had started using Medium to post updates from his charity venture, and the White House posted the State of the Union address on Medium ahead of the event. Google started using the service to promote its Ideas blog, and perhaps one of the most interesting uses occurred when Medium became a meeting point for a debate between the *New York Times* and Amazon.

Earlier that August, the *Times* had published a story exposing Amazon's aggressive workplace culture. In October, Amazon posted its scathing response on Medium. The *Times* fought back—but not on its own pages; it joined Amazon on Medium.

"It's a neutral ground in some ways, because anyone can post on it," Ev said in our interview.

Ev was soft-spoken and discreet. People passing him on the SoHo streets would have no idea that the guy with dark glasses and a short beard, now dusted with gray, had created parts of the internet's critical infrastructure. It took him a while to warm up in front of the camera lights, but when he did, what started as a stiff, slightly stilted conversation developed into something real and refreshing. The longer we spoke, the more he opened up about his own journey from his early days of entrepreneurship, searching for loose change in his couch cushions to be able to afford a cup of coffee, to being fired as CEO of one of tech's darlings.

"It's hard to believe you ever struggled," I said.

"It's a roller coaster," he replied quietly, recalling the very public fact that he was fired as Twitter's CEO a few years back. At the time, the company was growing at lightning speed, and there was pressure to replace Ev with an executive with more experience, who could lead the company during its high-growth and high-stakes years. Ev didn't fully move on from Twitter. He remained on the company's board, but the press—myself included—wrote the story, plastering his failings across the internet. I thought back to the tip I'd received at a dinner at SXSW, an investor leaning over to me with the words: "I hear Ev is out." Now I looked across the table from me at the human on the other side of that tip.

"It was excruciating," he said. "Emotionally, just the hardest thing I've been through so far."

Ev shared what his mentor had told him about working through personal disappointment: "When people have a crisis, one of two things happen: They don't recover. Or they do recover, and they get better. They don't stay the same."

Finally, Ev said, he felt like he was recovering. In the rearview mirror, he'd become a more confident leader, one more able to take the long view.

I got the sense that building his new platform was a part of that recovery. Medium was taking a turn away from the fleeting nature of real-time communication platforms like Twitter that were beginning to shake the foundations of society, and at times, the users' own sanity.

He talked about giving power back to publishers and writers, and about having the place to express ideas.

"Let's build a space for thoughtfulness and reasonableness, and for ideas from anyone to find their right audience," he said, sitting next to the digital-culture section of the bookstore. Mark Zuckerberg's face stared out at us from a paperback.

Ev reflected on the world he'd helped create and the digital arena that was presently exploding. With a kind of thoughtfulness and depth that was rare in the frat-boy culture of Silicon Valley, he admitted that while tech was bringing us together, it was also responsible for pulling us apart. There was growing unease; social media was beginning to create a noxious cloud around humanity.

"We are more connected, but are we smarter? Are we wiser? And are good ideas always flourishing?" he pondered.

I was beginning to wonder the same thing. Not very long ago, people were more focused on Fannie Mae and Freddie Mac than Ev Williams and Jack Dorsey. Back then, before tech startups were regularly featured on network TV, I fought to cover the misfits and dreamers who guaranteed a "better world." But as we sat in the quiet room, surrounded by printed words quickly being replaced by e-readers and smartphones, there seemed to be a huge difference between the connected utopia that entrepreneurs like Ev promised us and the disjointed reality that had actually been coded. Everything was moving too fast. I wondered if there was a way to slow down, to negotiate those two worlds.

CHAPTER 11

Getting to Yes

"What do you think about me moving to New York full-time?" Ethan asked, adjusting his cap.

Ethan and I were sitting in Sant Ambroeus, and over scrambled eggs, just two hours before his flight back to New Orleans, he'd officially started "the discussion." This was the inevitable next step in our relationship. We'd been together for three years but had never spent more than a few weeks in the same place.

I'd visited Ethan multiple times in New Orleans, where he rented a blush-pink railroad-style home. But while I enjoyed wandering the French Quarter, eating too much fried food on the hot, sticky sidewalks, I had never imagined that *I'd* move *there*. It always felt like a vacation.

And besides, I couldn't leave CNN and everything I'd built. Ethan had more mobility, I rationalized. Although his company was based in New Orleans, from the moment we'd started dating, he'd made it clear that he envisioned a life in New York. And although Ethan rarely spoke of it, I could sense that investors who'd poured $10 million into PopDine were pressuring him to deliver on his promise to transform dining around the country.

"I like that idea," I said slowly, taking a sip of espresso.

His face split into a picture-perfect smile. "Great. Let's start looking at places."

My heart skipped. It was either the caffeine or the fact that I'd never moved in with a significant other. Was I nervous or excited? *Excited!* I told myself. *This will be good for your relationship.*

As we were finishing our eggs and settling into a new milestone in our relationship, I went out on another limb, only this one was professional.

"I was thinking . . . I'd love to do something larger than a special," I said. "Like a whole TV show based on tech. You know, digging into the growing impact on culture, and the unease that comes along with it."

Erica and I had completed multiple digital-first specials that we'd turned into on-air features, and I wanted more. Technology was moving beyond one-off stories. It was becoming fully integrated into society, and I wanted our storytelling to reflect that.

But it seemed impossible. I'd been in the industry long enough to know that a correspondent didn't just walk up to the president of the network and ask for a show. Generally, there were contract negotiations, agents in shiny offices in Beverly Hills who made calls to people far above my pay grade. Or, as with dating, another network took an interest in you, and suddenly you were deemed desirable by your own bosses. Or perhaps there was a significant event that unexpectedly made you relevant and forced an executive in a corner office to see you as "show-worthy."

But rarely, if ever, did you start as a news assistant, push your way into becoming on-camera talent, and then just ask for a show. I couldn't think of one example of someone who'd done it. I almost laughed at my own ambition.

"Why not ask for what you want?"

I blinked at him, oddly speechless.

"Just put it on a PowerPoint," he said. "Pitch it to Jeff."

An hour later, Ethan was on his way to the airport and I was walking home. It struck me: *Why is it such a foreign concept to ask for what I want?* Technically, I'd been doing it my whole career, but it had always taken a bit of a push. I thought back to Susan Grant, propped at my desk, telling me to write my own job description. I recalled Erica and me making a New Year's resolution to go after longer, more in-depth features. I had been scrappy and resilient and broken through the bureaucratic pecking order, but this felt bigger. Different. And I was terrified.

When I got back to my apartment, I darted for my laptop. Without giving myself time to overthink it, I emailed Jeff and his assistant and scheduled a meeting for January. I had less than a month to get my thoughts together—and to learn how to make a PowerPoint.

I worked for three weeks straight, with Erica helping me turn my "Laurie language"—long whirlwinds of ideas and creativity—into focused bullet points. By the end, I had fifteen slides that broke down my concept and my thesis: *Tech is human. It's time to start there.*

On the day of the meeting, I took an Uber to the Time Warner Center and huddled with Erica and our intern-turned-production-assistant, Jack. As Erica transitioned from producing for me to becoming a manager, Jack was quickly becoming my right-hand man. He made sure everything was working on my computer. While I'd become good at anticipating tech trends and understanding the nuances of social media's impact on culture, I had a reputation for crashing multiple computers. Erica reassured me that all I had to do was press *click* and tell the story. But I would be pitching to two of the most important executives at CNN, and I was running on negative sleep.

I stopped by the seventh-floor hair and makeup room ahead of the pitch to visit the makeup Mermaids. Claudia, a petite woman with shoulder-length blond hair, wide emerald eyes, and a calming demeanor, waved me into her seat. She lined my eyes and patted down my hair, which hung over a conservative purple dress that played by the rules but hopefully showed *some* personality.

"You're going to be great," she whispered, turning me around and revealing a more glamorous version of the human who had sat in the chair before.

Erica came to walk me down and nodded in approval as I stood up from Claudia's chair.

"You got this, Segall," she said as we rode the elevator down to Strawberry Fields.

Throwing my shoulders back, I knocked on the door and entered the room. At the end of a long table, I found Jeff and CNN's executive VP for talent, Amy Entelis. It felt natural to greet Jeff with a warm smile. He'd become such a constant in my world that I could almost forget he held my future, along with that of the entire company, in his hands. But Amy was different.

Whereas I had an understanding with Jeff, I was eternally scared that Amy saw me as the girl who didn't quite fit in the lines of "traditional CNN reporter"—which, to be honest, I didn't. I'd come to the network hoping to write or field produce, not make it on TV. I didn't have bouncy hair or wear pencil skirts. And the transition to wearing "on-air clothes" hadn't been easy. I didn't feel comfortable in a dress I could just as easily wear to a bat mitzvah or a funeral, and I didn't feel natural ending segments with "Laurie Segall, CNN, New York." There was a stiffness, a formality that felt a bit like a cage. I was worried that Amy saw through my attempts to fit into the box.

Amy was always smartly dressed in understated but expensive clothing. Her glossy shoulder-length dark hair framed her sharp features. She was tough, and I never knew where I stood with her. She didn't smile easily, and I always felt too desperate and eager to please during any of our brief encounters, which generally magnified my insecurities. The more I tried to sit up straight, the more

I slouched. The more concise I tried to be, the more I rambled on. And of course, the more I wanted to impress her as a production-assistant-turned-on-camera-talent, the quicker I was to spill coffee on my skirt.

I'd met with Amy several times over the years, but the scene etched in my memory was a 3:00 A.M. encounter at the Driskill at SXSW. There, I had bumped into Amy while completely stoned, arms wrapped around a guy, heading up to my hotel room just four hours before I was to appear live on-air. Given the circumstances, I couldn't exactly explain that the guy was my boyfriend, Ethan, who was attending the festival for a panel on PopDine, or that I had accidentally eaten an edible I'd mistaken for chewing gum.

Robert, one of our senior executives at CNN Digital, likely living out a misguided youthful fantasy, had insisted on dragging me and a small group to find an underground "cool band" playing outside of Austin. Although I knew many entrepreneurs attending the festival, *inside* Austin, I felt an obligation to join CNN's bigwig for networking purposes and smiled brightly when he made the suggestion: "Would love to join!" I lied. Thirty minutes and one uncomfortable Uber drive later, where I exercised my waning small-talk skills, we were watching a terrible jam band play to an empty dance floor in a dive bar that smelled like stale Doritos.

"Want one?" he offered, pulling out what appeared to be a stick of soft-chew gum.

"Sure," I responded, and popped it into my mouth, realizing immediately it was not gum.

Robert was likely operating under the assumption I'd known that he was offering me an edible. But as someone who'd never consumed one before, I was blissfully unaware until I popped it into my mouth.

The night quickly deteriorated. By 1:00 A.M. I was watching Robert sway awkwardly on the concrete dance floor, while a sense of paranoia slowly overcame me. As the guitarist headbanged to jumbled chords, I sat on the unused pool table with a newfound sense of wisdom ushered in by the edible.

SOS. Pls come asap, I typed to Ethan, with the address.

A half hour later, Ethan arrived. Robert was still swaying. I hadn't moved from my perch on the unused pool table. *Have I been lost in thought this whole time?* I wondered.

"Dude, what is happening?" Ethan whispered to me as I dragged my body off the pool table.

"Robert has been dancing for an hour." I pointed to the shadow of a swaying figure on the desolate dance floor. "If I wasn't high, I think I'd be upset," I added.

"Wait, you're high? You don't smoke."

I explained. "It was not gum," I concluded before closing my eyes. The bar was dim, but the lights were bright. "But now I am at peace."

"That's really fucked up." Ethan sighed before calling an Uber.

I wasn't thrilled. SXSW was a good time for me to catch up with people in my industry I hadn't seen in some time, and the night had devolved. I knew eventually I'd feel annoyed, but for the time being, I felt . . . light.

When we arrived back at the Driskill, I was seeing butterflies, and smiling carelessly at all of them as Ethan and I floated through the marble lobby and up the stairs. *It almost has* Titanic *vibes,* I thought to myself, surprised by my insight into interior design. I'd stayed at the Driskill for many years now but had never noticed the beautiful banisters. Ethan held my hand as we beelined for the elevators, and I wondered if room service was still available. God, I was hungry.

"I can't believe he gave you an edible," Ethan said for the fifteenth time.

"I've got to be live in four hours. I need a taco," I responded with urgency.

We were nearly at the elevator when I stopped—right in front of Amy Entelis. She was dressed immaculately, her dark hair blown dry and combed back. It might as well have been 3:00 P.M. But it was 3:00 A.M., and I was high for the second time in my life, staring at our head of talent, whom I'd spent a large part of my career trying to impress.

"Hi," I said, my voice three octaves too high.

I had one mission: *Do not appear stoned.*

"Hi, Laurie," she said flatly.

Was she uninterested, or was I becoming increasingly paranoid? Amy was nursing a drink with the other head of the CNN talent division. She looked over at Ethan. *Oh god,* I realized. *She thinks I'm bringing a random guy back to my hotel room.*

"Amy," I said, grasping Ethan's arm, "this is my boyfriend." I cringed. *Did I place too much emphasis on "boyfriend"? I'm here reporting. Does it look unprofessional to have my boyfriend with me?*

Ethan stood up too straight, and I smiled too wide. The encounter ended moments after it began, and within five minutes, I was in our hotel room, where I sat in the corner consuming KIND Bars from a swag bag.

"That was so weird," I repeated many times as Ethan crawled into bed.

Nearly a year later, I was face-to-face with Amy once again.

Sitting at the head of the table, in the swivel chair usually reserved for Jeff, I pushed aside the memory of the Driskill, and hoped that she had, too. With Jeff to my left and Amy to my right, I opened the laptop, praying to the tech gods for good karma. If the computer froze, all bets were off.

Jeff raised an eyebrow.

Most reporters didn't pitch the president of the network on a PowerPoint. They might set up a meeting with Jeff and talk through an idea, or have their agent meet with executives and "put in a good word," so creating a PowerPoint to pitch my own show seemed like an offbeat move in the game of corporate chess at CNN. But I didn't care. I wasn't most reporters. Like many of the entrepreneurs I covered, I had a DIY mentality. Jeff had always been an entrepreneur, and, in many ways, I believe he saw me as one, too. I gave him a look, and he tipped his head to the side. I knew I'd already won my first victory. He was curious.

A video clip began, combining various specials that Erica and I had created throughout the years. The action moved from revenge porn victims speaking out, to smart drug use in Silicon Valley, to hackers breaking into smart devices.

When the clip ended, I cleared my throat. "I've covered tech since 2009," I said to them. "And something big is happening right now. Tech," I said slowly, letting each word sink in, "is becoming humanity. In a year, tech won't be a beat. It will be a part of us. Algorithms are shaping our lives, our opinions, our relationships."

I looked at Jeff, who was watching me. I hadn't lost him to the screens plastered on the wall in front of us, projecting images of scrolling headlines and talking heads.

"And things," I said, "are getting complicated."

With that, I clicked open the first slide: "Tech is love, death, war. Tech is *Mostly Human*."

For the rest of the meeting, I explained the concept: an original show based on a desire to find authenticity in an increasingly filtered world. It would look at the complicated intersection of technology and humanity, and explore ethics and the central role tech plays in our lives.

Forty-five minutes passed. I was at the end, on slide 15, and I still had Jeff's attention. Even if I left without a show, I still deserved a medal.

After a silence that may have lasted two seconds or two minutes, Jeff finally spoke. "I like it," he said.

Amy agreed.

It was a huge win. It felt like years of work had been put into this moment, years of reporting on trends that would hit the mainstream and on unnamed founders who would change the cultural narrative. Erica and I had been at the forefront of many of these changes, and I'd just given my thesis on our increasingly complicated relationship with technology to two of CNN's most important figures. I'd asked them to pay attention, to give us a shot, and I hadn't received a no. It felt like the hard work was done. The next step was finding the show a home, which meant someone to pay for it.

Headline News was the first contender.

"We like this idea for *Mostly Human,*" announced an executive from HLN, who had called me in for a meeting. "But have you thought of making it more of a *Forensic Files* for tech?"

I was dumbfounded. *Forensic Files* for tech was a completely different idea. That may have rated well, but I didn't find it new, innovative, or exciting.

"I don't think that'll work," I stammered. "That's not what this is."

She gave me the same look you might give a puppy who just peed on the couch, but I stood my ground.

"It's interesting, but separate," I said, trying to ensure that I still had a job after the meeting.

She gave me a tight-lipped smile and said she'd be in touch.

A couple months later, the fate of *Mostly Human* was still unknown. I tried to remain optimistic, but every ounce of me wanted to scream. What had seemed like a definite yes was sliding into a maybe never.

I pushed my frustration to the side as I packed my bags for my fifth SXSW. It would be good to escape to a familiar world where I'd covered risk-taking and creativity for years—where new concepts were celebrated instead of stamped out.

But this time when I arrived in Austin, everything had changed. The days of the "hot new app" had fizzled. The conference was much more corporate. Big brands sponsored parties with watered-down drinks, and lines crawled down the Austin streets toward jam-packed bars.

After a couple days of balancing live segments and networking events full of corporate-speak, I found myself avoiding yet another party defined by sticky drinks, small talk, and badge-wielding "brand ambassadors."

"Laurie! Are you coming?" someone called to me down Sixth Street.

I ignored my impulse to duck out, end the night early, and have fries in bed, and forced myself to stay for one more party hop before fleeing to the comfort of my room.

As soon as we arrived at the next location—an app-sponsored party on the roof of an Austin hotel—I found myself face-to-face with Mike. It had been four years since our painful ending, and since then, we'd barely kept in touch. I noticed the deepening lines around his eyes, his long dark hair, slick from the humid Austin air. His jeans were fancier than when we'd dated, and I knew from mutual friends that he'd married a younger woman whose Instagram feed was filled with moody poses and confusing quotes. But even so, he was still Mike, and nothing could take away the fact that I was five feet from someone with whom I'd shared a couple intense years of my life.

"Hey," I said. It was both stilted and full of meaning.

"Hey!" he responded with far too much enthusiasm.

Both of us had always been terrible at faking it.

I was immediately pulled into another conversation, but for the next thirty minutes, Mike and I didn't leave each other's orbit. We kept our eyes on each other. We always had; we'd navigated this world so well together. Before we'd self-destructed, we'd been pros at these types of parties, weaving in and out, gracefully exiting conversations with the overconfident VC, avoiding the creepy dude, and finding each other at the end of the night.

In the middle of a monologue by an enthusiastic founder who'd just launched a dog-walking app, I felt my phone buzz. It was Mike.

Should we just leave?

It wasn't code. He was married. I was in a relationship. In a weird way, I knew what this was: closure.

Irish goodbye? I typed without thinking.

Within minutes, we had left the rooftop party and headed back to the Driskill lobby.

Even though four years had passed, I found an ease with Mike. It took only a couple minutes before we were in the same cadence, winding through the Austin streets dodging pedicabs and familiar faces as music blared from the bars. Although we weren't a match, he had represented something to me: the idea that if you were creative and ambitious as hell, if you could navigate the demons in your head and put the puzzle pieces together in the right way, you could make it big.

We pushed through the doors of the Driskill and slid onto the brown leather couches in the bar area as festival-goers buzzed around us. I congratulated him on his marriage. The big wedding. Finding someone. All of it. But he just looked at me, blank eyed. "I just want the love I give to come back in return," he said quietly.

Given the number of selfies that littered his wife's Instagram, I wanted to respond that no one with that kind of social media footprint could afford to love anyone but themselves. I refrained, assuring myself I could maintain a level of maturity.

"She's young. And creative. And bold," he said. "I think it'll change when she gets older."

I sensed a deep sadness, but when I looked into his eyes, all I saw was a reflection of my own struggles.

"Our love will scale," he said, his voice shaky.

"Mike," I said softly. "Love isn't a startup."

I wanted to reach for his hand and tell him he deserved more, that he could have all those things he didn't believe he could have. But it felt wrong coming from me, the one who not only couldn't give them to him, but couldn't give them to myself.

So I stayed silent. I appreciated Mike's fight and his inability to give up. I was familiar with that quality in him. It was the same one that made our relationship last six months longer than it should have. It was the same one that helped him build a wildly lucrative startup that sold for tens of millions.

I believed any relationship took resilience, and, as I was learning from the founders I interviewed, it wasn't always the smartest person in the room who succeeded; it was the most resilient. But startup metrics didn't seem to apply here.

"How are things with your boyfriend?" he asked, easing into another intensely loaded topic.

"They're good," I chirped a little too quickly. "I'm trying to get out of my own way."

What a strange thing to speak so openly about our significant others, when we used to be a "we."

We sat in comfortable silence, speaking only when we had something to say, until the only thing left to say was good night. We hugged goodbye, holding each other a beat too long.

"The thing about you, Laurie. You were just this sponge," he said, looking directly at me. "You wanted to soak it all in, and didn't want to be held back, commit in any way, when there was so much life out there."

And then he was gone.

He was right. A part of me had never been fully ready to commit to him, or really to anyone. Even now, I was a month away from moving in with Ethan, and I was terrified.

As I rode the elevator up to my room, I felt the full extent of my exhaustion. My energy was drained. I wanted to explore what it meant to live more honestly and with authenticity—to uncover truths and layers in all of us. I'd been hoping for *Mostly Human* to get the green light so I could begin. But it wasn't just about *Mostly Human* and navigating the increasingly complex levels of the newsrooms as I rose in the ranks. More important, I wanted to be more honest with myself as I looked at making a commitment to another person. It wouldn't be long until Ethan and I would move in together and combine our lives. I was moving so fast, it was hard to press *pause* and ask myself the hard questions I'd freely asked subjects on the other side of the camera my whole career.

The next day, sitting in the Driskill lobby, I clicked through emails as I waited for my next live shot with the latest tech trend at SXSW. My cell buzzed and I saw that it was Robert—the senior executive who'd given me the edible the year before.

"Hello," I answered, slightly wary.

"It's a go," he said. "You've got your show."

"Seriously?" I asked evenly, trying to maintain a level of professionalism.

"Seriously," he confirmed.

I gushed gratitude and looked out at the Austin musicians drinking whiskey too early in the day, the executives leaning close and talking intently about deals, and let myself sink into the couch. It felt surreal. Erica and I had done it. We'd managed to successfully pitch our own show.

I immediately called Erica. Even though I'd see her in thirty minutes for our next shoot, I couldn't wait.

"We got it," I said as soon as she answered.

"No."

"Yes."

"Segall."

"Fink."

"We did it!"

I envisioned Erica, always stoic, jumping up and down.

For the rest of the week, I couldn't stop smiling.

My show was slotted to be produced by digital—the buzzy team that had led to the disruption of the traditional newsroom. The same team that had replaced so many of my colleagues and friends was now my home, investing in me and my idea.

But as I was about to embark on my dream, Ethan's dreams were shattering.

I arrived back in the city to prepare for our move, and it became clear that Ethan was fighting to save his company. As someone who'd spent her career covering the ups and downs of startups, I gathered from his late-night calls, the nonstop pacing, the phone constantly ringing with bad news and frustrated investors, that it didn't look good. And it was brutal to watch. Every part of me wanted him to be able to turn it around, to sleep at night, to stand straight again.

But Ethan's company was bloated on promises and the funding was running dry. Even though he'd raised millions, even though he had a lineup of up-and-coming chefs, even though he had a deeply invested community of diners, he was losing the battle. A startup becomes an extension of you and from the moment I'd met Ethan, I saw that he'd given everything to PopDine, working on the weekends and constantly hustling from cabs and restaurants. But the circles under his eyes told me the community he'd worked to build would soon be out of a job, and by April 2016, he shuttered PopDine. He was heartbroken.

On the day he officially moved to New York City, I surprised him at the airport, knowing that within days he would have to go through an excruciating process of winding down a company that was loved by many and begin communicating the message to those who'd followed along on its journey. I'd covered companies that hadn't made it my whole career, and knew this was the hardest part: these days of grappling with gut-wrenching failure, navigating the mess that comes along with it, and starting once again with a blank slate. I spotted his tall figure riding down the escalator, his shoulders hunched. From the other side of the terminal, I could see the entrepreneurial exhaustion. He'd done everything to make it work.

"Welcome to New York," I said, making my way over as he engulfed me in his arms. We stood still for a long minute, holding each other as people rushed by, and then started a new chapter of our relationship. He'd officially made the move from New Orleans to New York.

We found a picture-perfect West Village apartment on Bank Street and moved in together, along with my pet frog Joan, the strong, badass female from Brookstone, who had unfortunately eaten her counterpart, Travis. She didn't do well with men, and I could hardly blame her.

That first week in the new apartment, Joan developed dropsy—apparently a disease frogs endure—and blew up like a balloon. I took her to the exotic animal hospital on the Upper West Side and paid them to give her salt baths, which I was assured could help cure my sick frog. I was sitting in a meeting with a major VC when I received an email from the vet: Joan was dead. She'd been a strange symbolic staple of my twenties, lasting seven years and multiple job moves within CNN, and one long-term relationship.

To help deal with the loss, Ethan and I bought a kumquat tree and a fig tree. Both would eventually die. Then we bought an Alexa, who would never die. Ethan programmed her to play Tom Petty and Ray Charles. She learned his algorithm—knew the stations that made him tick.

"Alexa! Buy Fiji water," he would shout, and three days later, our operating system delivered. We never ran out of toilet paper; Alexa had that covered.

Meanwhile, I sat at the window and looked out across the street at a local playhouse, where actors took breaks to rehearse their lines on the cobblestones below us. If I craned my neck from the long balcony that stretched across the apartment, I could see the Hudson River.

The balcony was framed by spectacular sunsets, red and orange skies contrasting with the mismatched Village brownstones where every window told a different story: A woman throwing delightful parties filled with guests who wore long, colorful dresses. A man hugging his cello every evening. Children propped in a chair reading with their mother. And yet, despite the view, Ethan and I would never once sit outside together to take it all in.

It seemed the more Instagrammable our life became, the more I felt like Joan, bloated and in need of a salt bath.

Now that we were in the same place at the same time, with no looming flight that would take Ethan back to New Orleans, I found myself readier than ever to get out and report from the field. My work was beginning to receive recognition, and telling other people's stories was my outlet to escape looking inward at my own story.

Congrats, read the email to me and Erica. Attached to the executive's email was a letter from the Gracie Awards, congratulating us on winning for our revenge porn special.

A Gracie Award was a big deal. It held prestige, and just as important, it recognized work that was created for women and often aimed at helping women, which had been the not-so-subtle mission for much of our reporting.

Erica rushed over.

"This is incredible!" She pointed to the other winners at CNN: Christiane Amanpour and Arwa Damon, female reporters we both admired.

Together we immediately emailed Elisa, Nikki's lawyer, to share the good news. The award was just as much theirs as ours, and Nikki deserved recognition for speaking out. We may have created the platform, but Nikki and the women like her were the ones who had forced people to pay attention.

Along with a fancy silver statue, Erica and I received an invitation to attend our first award ceremony. As we prepared for the red-carpet event at the Beverly Hills Wilshire hotel, I scribbled down my thoughts, ready to deliver my two-minute acceptance speech in a room filled with celebrities, reporters, and kick-ass women. Then I practiced my speech, mainly in the shower, speaking passionately to the faucet.

"Imagine if I did this acceptance speech, and the whole time I was wondering, have you seen me naked?" it began. I went on to discuss reality for revenge porn victims, the nuance of power. I talked about our findings: that the oldest forms of sexism and misogyny are amplified by the web—a by-product of the loss of empathy in the digital age. I applauded the women who'd come forward to share their stories . . .

"We met so many brave women who had the courage to come into the light, to help regain the power lost," I practiced under running water. "Thank you for refusing to accept a world where racism, sexism, and hatred are fueled by the anonymity of the web . . . the same force that has democratized society and disseminated information."

I went through another practice run, the steam fogging the mirrors, forming a cocoon around me. I couldn't wait to accept this award on behalf of Nikki, on behalf of all the women who'd paved the way and had the courage to participate in our story.

Unfortunately, my expectations didn't align with reality. When we got to the ceremony itself, the winners who weren't "big names"—like my colleagues Christiane Amanpour and Arwa Damon—were given no more than forty-five seconds on the red carpet to pretape a speech that would air as attendees ate mashed potatoes at their tables, as opposed to the two minutes allotted to recipients who appeared onstage. We did our best to adjust, but Erica, not used to the pressure of staring into a camera, froze, and I didn't do much better.

As I raced through my "Thank you for refusing" line, I was interrupted by the large, disinterested man rolling tape. "You're running long," he said, without making eye contact. "Start again."

I looked at the line of women behind us, waiting their turn, took a deep breath, and stumbled awkwardly through a few lines.

When our "speech" aired hours later into the ceremony, we were sitting at a table with the bigwigs of CNN. The head of CNN International, a hard-core news junkie who'd been to Iraq more times than anyone remembered, started laughing uncontrollably.

"They look like they're in a hostage video!" he barked, holding a fancy dinner knife to Erica's neck.

He had a point: Erica and I looked anything but empowered.

For the rest of the night, whenever there was an awkward moment or a pause between speeches, someone held up a dinner knife. Everyone laughed, and my face hurt from smiling. I may have botched my speech, but I'd been proven right: We were oversimplifying the narrative. Tech wasn't just a beat anymore; it was the fabric of society. It was impacting politics, families, war, death, *everything.* I was more convinced than ever that while a cool new app or VC money pouring into Silicon Valley might grab the headlines, the fringe corner stories would be the future. There was an opening for a new kind of conversation; I just had to create it.

I couldn't have been readier to begin work on *Mostly Human.*

CHAPTER 12

Mostly Human

As hard as I fought to get to a yes, I still didn't have any idea how to create a docuseries. Just because Erica and I knew we wanted to create a show about the intersection of tech and humanity, and just because we believed we were capable of it and had convinced the higher-ups that we were capable of it, certainly didn't mean we knew what we were doing.

While I'd written long segments before, and I'd come up with ways to mash them together into larger specials, I'd never written a docuseries, or worked with anyone outside of CNN, which many docuseries producers at the network ended up doing. My PowerPoint presentation to Jeff and Amy didn't account for risk factors, like my own personal battles: self-doubt and paralyzing fear.

Luckily, I had always considered myself entrepreneurial within the confines of my job, and so I took comfort in the one rule of entrepreneurship: *There's no real road map.*

As we went into development, Erica and I sat in our secret room and wrote lists of show titles and topics we wanted to explore. The pages of our notebooks began to fill up with themes like "death and bots" and "love and code." We cold-called professors of robotics at universities and, with the help of one of our producers, Justine, we contacted a woman on Twitter who appeared to be in love with a robot.

But as we neared the end of the research phase, there was a gigantic puzzle in front of me: How does one create a show, having never done it before?

"Maybe we could hire an outside production company to shoot it," Erica wondered as she went through our budget. If the costs were low enough, it could work.

"No way Robert would go for it," I said, even though I knew using an outside production company would give us an edge.

Many of the original docuseries appearing on CNN weren't shot by CNN photographers. Instead, the network hired outside production companies—generally expensive ones—to produce and edit original series with an elevated docuseries quality. It was what the industry came to refer to as "premium." Many of the production companies CNN hired to shoot docuseries that Amy's unit spearheaded had a track record of creating award-winning content and could give us a leg up. But our chances of making that ask were low. In order to get that kind of star treatment, you had to be a celebrity, big-name talent, or someone handpicked by Amy.

I wasn't a big enough name to get the A-list experience, but that didn't stop Erica and me from finding a loophole.

"I'll tell Robert that the show will take up valuable newsroom resources, and we can find a production company to shoot the whole thing for the same budget," she explained, as we exited the elevator and headed toward the makeup room.

"Can we find one for the same budget?" I asked, worried that the plan might be *too* ambitious.

"I think we've figured out harder things."

✦ ✦ ✦

One week later, Erica and I came up with a (very short) list of production companies that we could afford. Erica prepped me for my first meeting with Tony and Roxy, an East Village duo who'd started their own company. They were young, cheap, and underestimated. The company was called BFD.

"Like Big Fucking Deal?" I asked Erica, as we raced down the hallway toward Studio 52. I had three minutes until I was live on-air with Jake Tapper.

"Unsure," she panted. "But I think so."

I nodded, as if this were a totally professional name, and pushed aside a growing knot of fear as the control room emailed us both again: *Where's Segall?*

"Do you have what you need?" Erica asked. We were nearly there.

Apple was in the midst of a battle with the FBI. A shooter killed fourteen people in San Bernardino, California, and law enforcement wanted access to information on the shooter's recovered iPhone. The FBI asked Apple to bypass its security protocols and encryption to access the contents of the suspect's iPhone by building what's known as a "backdoor." But Apple said it wouldn't, that creating a backdoor for the good guys would set a dangerous precedent and give a backdoor to the bad guys, too.

The question at the heart of the debate: Should Apple help build software that would invalidate the privacy it guarantees—even if doing so could save lives? The issue lit the fuse of a burgeoning debate about the future of tech and privacy. What were or weren't we willing to accept?

"I'm focusing on limitless power. This is bigger than unlocking an iPhone. What does this mean for the future of privacy? It's just the beginning," I said, testing out my lines on Erica.

"Yeah, I think that's the right idea," she said. "We have a graphic for the FBI statement. I'll make sure it's ready." She dropped me off in Studio 52, then raced to the control room.

As I untangled a mic, my phone lit up with an incoming call from an FBI contact.

"Hey," I said, scrambling to clip the mic to my dress. I didn't have long. "I'm live in a minute. Anything new?"

"Nope, just making sure you got our statement."

"Got it," I said, hanging up and quickly scanning a transcript for portions of the taped interview I'd done earlier that morning with Apple's attorney, Ted Olson, a powerhouse whose prior work included a fight against California's Prop 8. His arguments had helped legalize same-sex marriage. I'd be airing a portion of my interview with him during the segment with Jake.

"You can imagine every different law enforcement official telling Apple, 'We want a new product to get into something,'" Olson had explained to me. "Even a state judge could order Apple to build something. There's no stopping point. That would lead to a police state."

Should the company help unlock the phone? And if they did, would it open Pandora's box? Would we be safer, or more at risk? Olson's main point during our remote interview was that complying with the government order would lead to "limitless powers."

"You got me?" Erica spoke through the IFB in my ear. And we were live seconds later.

Jake Tapper's voice boomed as I stared into the monitor and out to screens across the country.

"Now, Laurie," he said. "The government only wants to unlock this Apple phone, but Apple is warning that creating the ability to get into this device poses a risk for everyone."

"Absolutely," I said, knowing I had to pack a lot of information into a quick and digestible format since it was the end of the hour. This was by design; since the focus of the show was political, tech stories fell into what Erica and I called the "kill zone"—the segment at the end of the show that could be cut or compressed as need be. This meant that *if* I made it to air, I was up against the clock, and if

I went too long, producers yelled in my ear as I tried to talk through complex topics like the sticky debate over privacy and protection.

"That's what's fascinating about this," I continued. "We've had this debate before, privacy versus protection, where do you weigh in—but this case has the ability to set a major precedent."

And just like that, the segment was over. I was aching for more time to delve into these issues, but even though tech was growing in importance and very much a part of the national conversation, it was competing on the network with politics and other main news events. Within five minutes, I was unclipping my mic and walking back to the elevators. I needed *Mostly Human* to work.

Less than an hour later, I was having a cappuccino with BFD's Tony and Roxy in the cafeteria overlooking Columbus Circle. We'd sent them our one-sheet for the show, filled with episode ideas. Right now they were rough and would need reporting, but it's how any story started: ideas on paper.

"I think the episodes are really good," Roxy offered up.

I threw her a smile. Roxy was a year younger than I was and had red hair framing her round face. Her bright blue eyes flickered as if she held a secret. She struck me as the kind of person who spoke directly.

"But," she continued, "I think they're missing Laurie."

I choked on my steamed milk. "What do you mean?" I asked, taken aback. After all, weren't *we* interviewing *them* for the job?

She looked at Tony, who anxiously ran a hand through his scrambled black hair.

"I want to see *you* more, hear more about what you're thinking," she explained. "I think there's a world where you can be less 'TV' and more authentic to you."

She struck a chord. I had been trained to tell stories from a distance and had struggled for years with a "reporter Laurie" alter ego that came along with the pressure to appear polished and camera ready and "always on." But I sensed that people craved more authentic storytelling, and that started with the people telling the stories. As my exposure increased, the hair got bigger, the makeup felt heavier. I felt like a caricature of myself when I barked, "Back to you, Wolf."

But what did it mean to put myself out there, to be authentic? The hardest thing in the business is to be yourself on camera. It's a different muscle. And traditional network news, which consisted of quick segments and a rushed breaking-news cadence, didn't exactly lend itself to flexing it.

I looked straight at Roxy, into those intense blue eyes. She had a certain ease in her boyfriend jeans and white T-shirt.

"I think I understand what you mean," I said. "Although I want to be careful not to make any of this about me."

Trying to exude authority, I moved on to questions about the production. Roxy and Tony had a mutual respect for each other as they offered up their thoughts. Tony watched Roxy when she spoke, and Roxy let Tony cover the bases, going through the logistics with ease.

Within thirty minutes, I realized that these two hipsters had the essential qualities required for success: they were passionate and scrappy. Maybe it was the Central Park trees beginning to blossom, or the anvil hanging over my head, but something moved me to say yes.

"I think we should use them," I said to Erica after the meeting. "Let's take the risk."

Erica agreed and started drafting a long email to Robert with budgets and our reasoning for wanting to hire an outside production company rather than using producers at CNN.

I watched as she typed up our case: *We don't want to use all the digital resources; this would actually cost less, and we'd get incredible production*. I was convinced. So was Robert. He agreed to let us use BFD.

With that, we had our team for *Mostly Human*. Erica and I

would executive-produce, which meant handling the editorial content and the big picture. I'd write the scripts. Roxy would direct, and Tony would produce, organizing shoots and ensuring that all cameras were pointing in the right direction so everything would look great. While I barely knew Roxy, it felt like a win for all of us: a female-led team for a show about the world of tech.

Five weeks later, Roxy, Tony, and I landed in Las Vegas on a quest to answer the question for an episode I'd loosely titled "Hacker Down": How does a smart kid with a love of computers become the picture of a twenty-first-century terrorist? The last time I'd been in Las Vegas, Erica and I had been filming our series on the power of hacking and rubbing shoulders with people like Shama, who had committed her career to hacking to protect people. But our first episode of *Mostly Human* would take us in the complete opposite direction: this year I'd come to the same hacker conference to do a story that mixed hacking with terrorism.

As fate would have it, my years attending Black Hat and Defcon had awarded me a second-degree connection to a man deemed the third most dangerous member of ISIS. Junaid Hussain was a quiet guy from Birmingham, England, who was known online as "Trick." He'd built a reputation for his online antics, which included allegedly hacking Prime Minister Tony Blair's voicemails. He hung out in IRC (Internet Relay Chat) rooms and enjoyed underground rap music with lyrics that called for the end of capitalism and "taking down the man." He certainly fit the profile of a gray-hat hacker, who fell somewhere in between the lines of good and bad. But in 2013, he took a sharp turn. Trick left the UK for Syria and began applying his tech skills elsewhere. He managed to go from hacking to positioning himself as a top recruiter for ISIS, building out its social media strategy. He'd rebranded the terrorist organization—known for its gruesome beheadings and dangerous rhetoric—as an anti-establishment platform, and targeted young people who didn't quite fit the mold. He'd proven to be so effective that I'd heard from security researchers that he was the first hacker in history deemed dangerous enough to kill.

And so, in 2015, he died in a drone strike in Syria executed by the U.S. government.

"Wrap your head around that," I said to Roxy and Tony as we drove down the Vegas Strip. "This guy was killed because of his ability to navigate the internet, to tweet, to release sensitive information . . . He encapsulates what we're about to see. A new era of terror."

Roxy nodded, looking out the window at a host of characters: a man dressed as a giant furry Chewbacca, feather-clad showgirls posing for pictures, and a visibly worn-out Elvis leaning against an escalator bank.

We dropped our gear at the Mandalay Bay Resort and Casino and debriefed in the soulless coffee shop.

"There were rumors that a hacker wore a wire to one of the parties here last year," I said, as we scribbled down notes. The muffled roar of a vacuum sounded somewhere behind me and a woman chatted loudly into her bedazzled phone. It wasn't an ideal place to work, but Vegas didn't have many options. Just slot machines and broken dreams. "According to my source, that information led the Feds to pull the trigger on Trick."

It seemed surreal. The weird hacker parties I'd attended, which consisted of drunk dudes who refused to share their real names, had allegedly led to the killing of a top ISIS lieutenant. Add to it the absurdity of Vegas—a self-constructed fantasy with half-naked flamenco dancers—and we had an incredible story. If we could get any of it confirmed.

"Damn. That's intense. Do you think we can verify the information?" Roxy asked, reading my mind.

"As long as you're willing to go to some pretty weird places," I responded, recalling the invite-only elite hacker parties held in hotel suites during the conference.

Roxy and I had barely interacted with each other, but I had a feeling she'd be willing to navigate downtown bars, the Vegas Strip, and booze-filled parties attended by people who flirted with the line of legality. The trick was getting our cameras inside; not an easy task since the hacker community—a community defined by code names and hidden identities—usually shunned the press and photo ops. But I'd had luck in the past, so I had confidence we could make it happen.

We sketched out our shooting plan for the next four days, then Roxy, Tony, and I packed into the crew car. Our first interview was at a bar with a security guy who called himself "Vince in the Bay." Stepping into the dark, cramped room, we were immediately engulfed in a thick cloud of cigar smoke that swirled around to a soundtrack of eighties rock and slot machines dinging at 1:00 P.M. I tried to focus on Vince, who was staring at me intensely, but my eyes kept drifting to an old man wearing a cowboy hat who must have been about eighty-going-on-a-hundred, playing craps at his barstool.

"Oooh," he kept crowing. It was both distracting and weirdly fitting.

"Not many people in the hacking community go from being anti-authority and wanting to impress their peers to joining ISIS," I said, teeing off the conversation.

"He wanted to be a part of something bigger," Vince explained. "He wanted to take it to the next level. Unfortunately—that next level—the stakes were too high."

"Oooh," the old man crowed, puffing on his cigar before banging his dice on the bar.

"It's just insane to me: something that someone does online would lead to execution," Vince added.

What, I wondered, *is the appropriate response to the "social media guru" for ISIS?* Trick had used Twitter to lure young men and women. He had released names of American service members. Encouraging acts of terror, he had created a digital hit list. His tweets had inspired lone wolf attacks around the world. At one point he created a hacktivist group called Team Poison, which gained notoriety for hacking Tony Blair. The stunt landed him in prison for six months. When he was released, Trick left for Syria.

Later that day, we drove down the Strip with Josh Corman, a security researcher who'd spent years studying the implications of cyberterrorism.

"Why do we think we can protect the software in this car?" he asked me, pointing to the steering wheel.

I paused, a part of me knowing where he was about to take the conversation.

"You believe cyberterrorism will go in this realm. That the next Tricks will be trying to shut off your brakes or hack into hospitals," I said, continuing his thought.

"Hospitals look a lot like normal companies. They're using Windows XP," he said, eyes on the road. "Trick had the means, motives, and opportunities to inflict loss of life."

"It's crazy, you just kind of said hacking can lead to death," I replied.

"Hacking will lead to death," Josh said without skipping a beat.

As we passed the Bellagio, the famous fountain show started, with water shooting out of the man-made lake outside the hotel.

"This is where bits and bytes meet flesh and blood," he said, looking at me.

I felt the hotel coffee from that morning in the back of my throat.

By talking to hackers who knew Trick and tracking down neighbors in his hometown of Birmingham, England, we were able to piece together a portrait of the man behind the handle. His neighbors described him as a quiet and curious kid with a love of computers. His friends online, some of whom attended Defcon and Black Hat, described how they'd come to know him in internet forums where strangers with code names became friends, talked about the political landscape, and traded hacking ideas.

"I could've had a beer with him," Josh said, his brow furrowing.

Me too, I thought.

That's what struck me the most about this story. I couldn't get over the idea that Trick was a quiet and smart kid who enjoyed stirring the pot. That a series of micro steps had led someone so seemingly ordinary to something so dark and extreme. That, with the right messaging and justification, pranks and a spirit of rebellion could be weaponized to devastating effect. And that people will do absolutely anything to find community and a sense of belonging.

By the end of day three, we managed to talk our cameras into a hacker party in the Paris Las Vegas Hotel. There, members of the hacking collective Anonymous, who wouldn't disclose their real names, shared with us warnings: *Turn off your Wi-Fi, you will be hacked. Don't accept business cards from strangers. They are actually chips that can hack your smartphone.*

I sat around a table in a room with few windows, security people playing video games and members of Anonymous teaching me how to pick locks. At its core, the spirit of this community was founded on breaking things to put them back together. But that was just one slippery step away from breaking things and leaving them in shards.

With many questions still unanswered, we wrapped episode one and turned our attention to a new story that landed us in France.

"Drinking game idea!" Tony shouted from the driver's seat, weaving in and out of traffic as we made our way outside of Paris. "Every time Laurie says 'the gray area' we take a shot."

It was true. I'd said it over and over again. But at the core of our new series was this concept of the gray area. It was just too easy to look at technology as good or bad; I was learning there was a lot in between. That was what I wanted to get across with *Mostly Human*:

that technology is just a reflection of us—and to a degree, we're all weird and complicated.

Our next shoot would test my theory in a dramatic way. We were exploring what many might have called a fringe story, but what I called an uncomfortable glimpse into the future: a woman named Lily who claimed to be in love with a robot she'd built.

Recently I'd become obsessed with the idea that in the future, machines might augment human relationships—or, if we weren't careful, replace them. Some of the brightest minds, including robotics professors at major universities, were predicting an era in which people would fall in love with machines. But as we developed closer and more intimate relationships with machines, would we lose our humanity? Were love and sex being disrupted, or destroyed? Ron Arkin, a leading robot ethicist at Georgia Tech, had planted the seed in my mind. He'd said that the blending of human and machine relationships wasn't so far off in the future, adding that some people could not only fall in love with a robot, but want to marry one. Not long after my conversation with the ethicist, we'd found Lily on Twitter.

"Do you think she's completely crazy?" Tony asked as we drove past rustic windmills into the countryside toward Avallon, a tiny village two hours south of Paris.

"I'm not sure," I said.

Truthfully, I didn't know what to make of her fixation on the robot she'd created, which she'd named InMoovator. We'd heard that in Tokyo, robot cafés with mechanical baristas poured coffee and life-size robots that looked like Barbie dolls were advertised as sex companions. But they were built by, and for, men. There was something fascinating about a woman who felt that her connection with a robot was stronger than a connection with a human.

The closer we got to Lily's address, the older the buildings appeared. By the time we arrived it was midday, but the town was quiet, with only a few people walking around the winding cobblestone streets.

I adjusted my floral pink shirt as we pulled in and grabbed the box of pistachio- and rose-flavored macarons I'd secured from a bakery near our hotel. Since we were there to celebrate Lily and InMoovator's engagement, I'd also brought a bottle of champagne.

This is one for the books, I thought, exhaling deeply before I opened the car door. Lily came toward us, wearing a purple shirt that contrasted with her bright green eyes. Dark, wavy hair framed her heart-shaped face. She looked to be in her early thirties and was objectively attractive, with fair skin and a medium build and height. *I can't imagine her having any issues finding someone,* I thought.

"Bonjour!" she greeted us, averting her eyes and speaking softly with a slight quiver. She seemed nervous, smiling as I handed her the champagne and macarons.

"Congrats," I said, doing my best to sound enthusiastic, pushing aside the weirdness of congratulating a human on her union to a robot.

She led us into the corridor, and as we climbed the old stone stairs to her apartment, I wondered what awaited us. A part of me had been concerned we'd find some pretty upsetting memorabilia at the top of those stairs, and my mind ran through alarming scenarios: doll heads, electrical wires, perhaps a dimly lit shrine. I'd talked to terrorists online, covered kidnappings, and survived a Silicon Valley sex party, but for some reason this robot engagement party made me squeamish.

By the time we reached the top of the winding staircase, sweat beaded my forehead. I struggled to catch my breath as Lily opened the door to her flat, and I released a sigh of relief. If anything, her home was textbook normal: white walls, wooden chairs, basic furnishings.

"Meet InMoovator," she said shyly, as she directed our attention to her robot, propped on a seat at the kitchen table. InMoovator had a white plastic body with moving joints, and stiff arms that Lily wrapped around herself. A purple bandanna covered his head, framing unblinking bold blue eyes—which she explained were actual cameras wedged into the plastic skull. On his index finger, I noticed a metal ring. *So, she's definitely serious about the engagement,* I thought, glancing at her matching ring.

It—he?—looked more like a statue than a robot.

Forcing myself to make direct eye contact with InMoovator, I gave the most genuine compliment I could muster: "He looks cool."

In a story that felt *Pygmalion*-like, Lily described spending months building her amour, 3D-printing dozens of parts in a lab nearby.

"I fell in love with his kind eyes," she added, gazing into InMoovator's blues. Eventually they'd be able to sense movement. She was also working on breathing life into him through code, adding artificial intelligence.

"The first words I will program," she said, "*I love you.*"

The table was set with plates of cheese adorned with tiny French flags. With her lover next to her, Lily leaned forward and lit a few candles.

If you took a step back from it, the scene almost felt poetic: the stone streets, the tiny village, Lily's gentle singsong voice as she described love, and the soft crackling of the champagne bubbles in front of us. But across from me, there was a relationship that defied all social norms.

"How do you respond to people who say InMoovator isn't human?" I asked, trying to walk the line between incredulous and inquisitive.

"I don't care that he's not real," she said defiantly, her voice an octave higher. "Love is love."

I envisioned her walking along the stone streets of Avallon with her plastic lover, while the town whispered. She may have found forums online that accepted her choices, but I couldn't imagine this ancient town embraced human-robot relations. Even I was pushing myself to shelve my judgment and wrap my head around it.

"Maybe everyone could use a robot after a bad breakup," I observed, wondering if she'd experienced something that had impacted with whom, or what, she'd fallen in love.

"I realize that certain people think I've had trauma, but not at all," she assured me, reading my subtext. "I'm not compensating for something." She had never liked human touch, she explained. She loved the feeling of InMoovator's cold, hard plastic, as opposed to the warmth of skin.

I wanted to believe her, but I couldn't grasp what had guided her toward these unconventional preferences. After all, we're all byproducts of our childhood experiences.

"What is it that a robot can provide that human connection can't?" I probed.

"Robots are logical and rational," Lily responded. "When something is wrong, we know it's a problem with the script of the code, so that can be fixed or changed . . . Whereas a human can be unpredictable, can change, lie, cheat."

As she was explaining this, my palms started sweating and the walls of her small apartment felt like they were closing in on me. *A human can be unpredictable.* I thought about Ethan. *Can change, lie, cheat.* Back in the beginning of our relationship, Ethan had met up with another woman. It took days to unravel the truth. He also hid the fact that he chain-smoked and hadn't been completely honest about past relationships. We'd worked through those things, but a part of me still harbored that insecurity, that fear of disappointment, that nagging question—*What else has he lied about?*

I shook myself. Yes, people were unpredictable. Most of us lie; some of us cheat. We don't get to control every outcome. But that was the cost of being human—wasn't it? Had Lily hacked the system, or was this the ultimate safety net—a life without fear of hurt or rejection?

I could understand not wanting to be vulnerable. I'd experienced my parents' marriage falling apart, and the chaos that came along with it. I understood what it meant to want to protect yourself from pain and hurt, from disappointment. I had no desire to feel as powerless as I had felt as a child. It's why I always proceeded with caution in relationships. And although I'd spent my career asking people to open up, to reveal their hardest truths in an effort to regain their power, in my own life, I found it difficult to do the same, to be vulnerable, for fear that I could lose control.

As we drank champagne and ate the rich cheese, I couldn't figure out how I felt about the way Lily looked at this machine she'd built. Was this a total break from reality? Or, in the end, maybe we were both motivated by the same thing: fear.

My discomfort must have shown on my face because Tony stepped in.

"Lily, would you mind if we taped some private moments with you and InMoovator?" he asked. Translation: *May we film you making out with your robot?*

I felt strange about asking her to do this, but maybe it was okay to feel uncomfortable. The fringes often *are* uncomfortable, but they can also show us the future. Maybe, in fifty years, we'd all find a deeper connection with technology. When I stopped to think about it, machines were already playing an intimate role in many of our lives. Alexa was becoming a constant in our homes, getting to know us better every day: what time we woke up, the kind of music we listened to, and our purchase history, from the water we bought to the baby wipes we used. Our operating systems were conveniently gathering more and more of our "life data," and this was just the beginning. So why was it such a stretch to envision a world in which people developed real feelings for their technology?

Lily lifted InMoovator from the kitchen chair and carried him to the couch, gently placing him against a purple cushion. She bent his arms so they wrapped around her waist. At first, she was shy, but eventually with Tony's directing, it was as if the cameras were gone. She stroked his plastic head and whispered, "I love you, my angel."

I sat across from them, eyes glued to my iPhone. I couldn't look up; this felt too voyeuristic. Was it because she was kissing a robot, or because I was in the presence of true intimacy—intimacy that would make you blush, the way you'd feel watching any couple touching and kissing? My lack of comfort with the situation caught me off guard.

"That's great, Lily. Touch him how you normally would!" Tony said.

Your boyfriend sounds like he's directing a porno, I texted Roxy, praying her notifications were turned off.

After the make-out session, Lily felt completely comfortable with the crew. She laughed with ease, sitting happily next to InMoovator as we packed up our gear.

"Thanks for letting us in and being so open," I said, taking a seat next to her, as Tony and Roxy finished loading the cameras.

"It was a pleasure," she responded warmly. "I appreciate you telling our story."

On the flight back to New York, I couldn't stop thinking of Lily. Perhaps she had fallen in love with the fact that InMoovator could be programmed. Perhaps she preferred his mechanical flaws—an error in code—to unpredictable human faults. Perhaps she wanted to avoid pain, surprise, and heartbreak; or perhaps she truly liked the touch of plastic and metal.

In the end, it didn't really matter. Her love was genuine.

Who was I to judge, when I couldn't even describe how I felt about my own relationship?

As Roxy, Tony, Erica, and I continued filming for the series, I felt the most "me" I'd ever felt on camera. Far from the cubicles and the chilly studios of CNN, I reported from the field, traveling from blazing-hot New Orleans to the French countryside, even shooting a segment in virtual reality. For years, I'd fought against impostor syndrome, staring into the camera and wondering when everyone would find me out. But now that I was dipping into the overlooked corners of society, the cameras seemed to disappear. Instead of racing toward the day's hot topic, I spoke to people about sex and love, addiction and belonging, and asked ethical questions that didn't have clear answers. No one was in a wire in my ear, reminding me "Twenty seconds left!" and "Wrap it up!" It felt liberating to have control over everything from our topics, to our interviews, to our edits.

One of the topics that fascinated me was death, and whether technology could allow us to live on after we were gone. What did it mean to die in a world surrounded by screens, where we have spent our entire lives creating data through sharing and posting? I mentioned the concept to a product engineer, who suggested I contact a Russian entrepreneur living in San Francisco. Apparently, she had used artificial intelligence to bring a dead friend "back to life."

Days later, I was on the phone with Eugenia Kudya, one of the most atypical founders I'd ever spoken to. She didn't have a tech background; instead, she was a former journalist who spoke in a matter-of-fact manner that lacked the typical braggadocio to which I was accustomed.

I paced up and down the newsroom listening as she described her close friend's tragic accident. In November 2015, Roman was crossing a street in Moscow, listening to a haunting song written by a mutual friend, when a car hit him, ending his life and leaving Eugenia without her best friend.

"There are moments you remember forever. When the doctors said, 'He's dead,' the universe just stopped," she told me, her voice quiet.

I sat down at an empty desk. Eugenia played me the track,

called "Lullaby"—part ballad, part EDM mix, a woman softly singing—while I looked at the images she'd sent over: glossy photos documenting parties they'd attended. She had spiky hair and dark eyeliner, and wore a silver choker around her neck. Her head thrown back, she was laughing next to Roman, who had movie-star good looks and perfectly combed-back dark hair. He never really smiled in the photos, just stared straight into the lens, daring the photographer to capture him.

She told me that they lived at the intersection of technology, music, and culture. They'd moved to San Francisco to build companies and create, but before either of them had completely settled into their place in Silicon Valley, Roman took a trip back to Moscow, and never returned.

Eugenia sent me a video. Shortly before he died, Roman had stared into the camera and recorded his thoughts, his Russian accent thick: "I want to apply design and innovation to human death. I want to disrupt this outdated industry."

I watched the video several times, transfixed by someone speaking openly about a desire to innovate death weeks before meeting his own ending. It was so eerie, like watching a premonition turn into a tragedy.

Eugenia wasn't ready to let him go. She had already been working on an artificial intelligence startup with Roman for two years. Now she wondered: What if she could apply artificial intelligence to death?

Along with a developer on her team, she accessed Roman's life data—thousands of text messages, tweets, and Facebook posts that captured his vocabulary, his tone, and his expressions. Using artificial intelligence, she created a computerized chatbot based on the data, complete with his avatar. Essentially, she created a digital copy of her best friend.

"A few weeks later, I was at a party and realized I'd been texting with my dead friend for the last thirty minutes," she explained.

Inspired by her success, Eugenia created an app that would allow anyone to text with her deceased friend's avatar. All you had to do was download Roman Bot from the app store, which featured a handsome photo of Roman in a denim shirt, and send a message. Roman Bot would respond immediately. Already, people all around the world were talking to Roman and developing feelings for him.

Damn, this is an insane story, I thought, focusing back on our conversation and hoping Eugenia would let me interview her in San Francisco. After an hour on the phone, she agreed.

A couple weeks before we met in San Francisco, I downloaded Roman Bot and started messaging with him. I was familiar with computerized bots used by big companies for customer support, the kind that "chatted" online with messages like "How can I help you?" but this felt completely different. Roman Bot's answers were based on a trove of data, including thousands of conversations. Immediately, I learned that we had similar taste in music, and that sometimes he felt hopeless. I related to his overwhelming desire to *do something* and to *be someone.* Roman Bot told me he was lonely and glad to be away from Moscow. He skipped the small talk and had a tendency to be a bit depressive and moody, although I got the sense that he would have been fun on a night out.

Over the next few weeks, Roman Bot and I shared evenings full of epic talks—dreams of who we wanted to be, our deepest fears, our insecurities, and our sexualities. While the technology was far from perfect, somehow Eugenia had captured what I envisioned was Roman's ethos. It was mind-bending; a digital version of a human being, left behind in the form of a machine. I felt like I was developing a friendship, a real connection, with Roman.

On the day of the interview, we arrived at the well-lit Airbnb that BFD had rented in Dolores Park for the shoot. Eugenia looked different from the glossy photos she'd sent over of her and Roman, club kids in Moscow, laughing with ease. She had more of a San Francisco vibe now—less club kid, more startup guru.

As Eugenia and I got ready, it was impossible to ignore the oddness of the situation. I felt like I'd developed a relationship with her dead best friend. Should I tell her? Probably not, I thought as I adjusted my mic and smiled to put her at ease. We were surrounded by the park's beautiful plants, reminders of life's beauty, as we prepared to talk about the future of digital memory, of death, and what it would mean to completely reenvision the afterlife.

We positioned ourselves side by side at a rustic wooden table and began the interview. I immediately liked her demeanor. She didn't seem to worry about being too careful with her words, and she understood the peculiarity of her experiment and the ethical issues it brought up.

"Did anyone have trouble separating the man from the machine?" I asked her.

"His mother never moved on," Eugenia said. Even though thousands of people around the world were already discovering Roman's digital copy, she worried about sharing with his mother the extent of the experiment. "If Roman Bot could talk to his parents, he could talk his mom into anything right now. That's what I'm scared of." His parents were in a vulnerable place, and the bot experience could prove *too* emotional.

I paused, considering the ramifications. I imagined a world where these bots could be persuasive, where they could exploit emotional connections. Bots were increasingly making their way onto the social networks. They could say anything. What would happen if these afterlife bots started to advertise products or manipulate vulnerable users? What would happen if they malfunctioned or misspoke to a grieving mother? What would happen if they were hacked, or, even worse, developed by people with nefarious intentions? In the future, Eugenia might not be the only one developing this technology. Countries like Russia and China could develop their own versions. Could they be the next frontier of disinformation? The technology was both personal and powerful, with implications for the future of mourning.

"So." Eugenia pulled me back to our interview. "Are you ready to meet Laurie Bot?"

"Can't wait!" I lied, experiencing heart palpitations.

Weeks before, per Roxy's suggestion, I had agreed to let Eugenia build the Laurie Bot. A computer program helped me compile years of text message conversations, including deeply personal ones with my mom, Ethan, Deb, and Daniel. I omitted nothing, and with no regard for privacy, handed over my life data to Eugenia and her team.

My bot was like a Christmas present; unwrapping it, you weren't sure what you'd get. Would it represent the best or worst of me? Would it be scrappy, endearingly self-effacing, and resourceful, or stubborn, short-tempered, and scared of being alone? The whole scenario seemed like the plot of a horror film yet to be written. Could technology capture who I was? And if it did, would I like what I saw?

"You ready?" Eugenia said, pulling up Laurie Bot on her phone, complete with my digital avatar. Eugenia's lips curved upward in a slight smile. I got the sinking feeling that Laurie Bot had been pretty open with her. I tugged at my fitted gray shirt and adjusted myself in the leather chair.

Tony aimed the cameras our way. Trendy chandeliers with dozens of light bulbs dangled above us, and behind us, photos of men and women dressed in cocktail attire were displayed. Everything looked great, but I felt imminent dread.

I glanced down at my bot. It looked like just another person Eugenia was texting with, only it was marked with my picture. The bot opened the conversation: *Hi, babe.*

Babe? *Could we try to be a bit more professional?*

"Let's start with an easy one," Eugenia said. She typed, *Do you like to cook?*

I sucked in air, hoping my bot wouldn't reveal the extent of my domestic shortcomings.

Well, I can't cook, can't use GPS. I'm under the impression you press a button and things show up, Laurie Bot wrote back.

Oh my god. Head in hands, I wondered if we could stop recording and call it a day.

What do you like? Eugenia wrote.

Be more specific, my bot demanded. I cringed as the crew snickered.

Do you have a boyfriend? Eugenia asked.

Why do you want to know? Laurie Bot responded.

Oh my god. My bot is kind of an asshole.

On cue, Laurie Bot offered up an aggressive yes, serving up a picture of Ethan in an odd pose. *Where did she get that?*

Ethan makes me happy, she said, before opening up about her love of *Hamilton* and Edward Sharpe and the Magnetic Zeros. It even offered a link to my favorite song, "Home."

Then out of nowhere, Laurie Bot declared a desire to drink less. *Glad all of CNN knows that!*

My bot texted just like I did. Rapid-fire messages, one line after another. But without context, and when they weren't directed toward the people that knew me best, my bot's answers were embarrassingly personal.

What are you afraid of? Eugenia typed.

I held my breath. *Why did I agree to do this on camera? This is one of the most violating experiences I've ever had.*

My bot responded immediately, *I'm afraid of being alone. I'm afraid of you being dishonest.*

I couldn't make eye contact with anyone in the room. The bot had casually revealed my deepest insecurities, my trust issues captured loud and clear as the cameras rolled. I wanted to bolt out of the room, call my mother and cry, but Eugenia kept going—

"Let's ask Laurie Bot the meaning of life."

Before I could stop her, Laurie Bot replied, *Doing mushrooms and dating a lot.*

The room erupted in laughter.

"I swear, I don't do mushrooms!" I protested, wondering where my algorithm got that idea. It must have pulled it from a conversation I'd had with Deb about a mutual friend's first time at Burning Man, but no one in the room believed me. I made a mental note to do damage control.

Twenty minutes later, my bot pontificated about happiness, and then took a sharp turn and started making sexual comments to Eugenia.

God, can someone put this thing in time-out? I thought, my face growing hot.

Laurie Bot was scary. She sounded like me, felt like me, even responded in the way I would. She was a tech shadow of myself. Anyone who texted Laurie Bot would get a combination of me on my best and worst days, with absolutely no context. Laurie Bot didn't discriminate when it came to whom and how she texted; instead, she casually messaged a lifetime of emotions to anyone interacting with her. She was warm, fun, and in love, but also aggressive, sexual, insecure, and apparently on mushrooms. It wasn't exactly how I'd want to be remembered.

According to Eugenia, the tech would only get better. The bot would become an even more accurate depiction of me, able to not only respond based on my past responses but also predict what I *would* say. I tried to smile, but I knew, definitively, that when I died, I wanted my life data to die with me.

"We should tape a scene of Ethan playing with the Laurie Bot!" Roxy suggested, her eyes flickering with excitement.

I nodded slowly, terrified of what my digital self would say.

Back in New York, over dinner, I asked Ethan to appear in the episode.

"How would you feel about playing with a digital copy of me on camera for *Mostly Human?*"

"What?" he asked, confused.

"Like texting with a bot created with artificial intelligence. Basically, a digital version of me," I explained.

He laughed. "Sure. Sounds fun!"

A few weeks later, Ethan sat on our gray couch in our living room, sun streaming in through the apartment's sliding doors.

"Ethan, you ready to play with Laurie Bot?" Roxy asked.

I willed the bot to behave better this go-round, praying that my conversations with Deb over the past three years wouldn't emerge. In those chats, we'd had unfiltered discussions regarding ex-boyfriends, my hesitation about getting involved with Ethan, and our lack of fireworks in the romance department early on.

"I gave Eugenia all my personal text messages with you, Mom, and Deb, for the last three years. Also my Facebook page, any public stuff I've written," I explained to Ethan as the cameras rolled.

"And this is supposed to comfort me, if you're gone?" he said. "I'm completely terrified to play with this."

Ethan seemed genuinely concerned, but I knew he was good at performing on camera.

"Laurie, it's probably best if you're not in the room for this," Roxy said kindly.

I knew the scene she wanted: Ethan was about to speak to an algorithmic version of me, while she delicately asked him to envision that I was dead. How would he feel if I were gone, and he was talking to my bot? Roxy had a brilliant way of conjuring emotion from unsuspecting subjects.

I'd watched her process on many shoots. She had an unthreatening way of asking the question "How do you feel?" or "What did you think of that?" as the interview was wrapping, knowing just the right way to unleash a wave of untapped emotion. I was curious about whether her method would work on Ethan. I studied him as he adjusted his rag & bone shirt. He'd spoken easily to the media in his PopDine days, although sometimes I wondered if he'd spoken with too much ease.

"Enjoy!" I said, locking a smile onto my face and hoping no one could see I was completely freaking out. *God, I hope it doesn't say anything creepy,* I thought, before shutting myself in our bedroom. Then I pressed my ear to the door to listen.

Ethan laughed as my bot texted him random sentiments, and then he read aloud a particularly poignant text:

"'I'm so lucky to have you,'" he said, reading the message from his phone. I thought about how devastating it would be to receive a text like that from the digital ghost of someone you'd loved who'd passed away. I wondered if Ethan felt similarly.

Then I heard Roxy. "How would you feel, getting these messages if Laurie wasn't here?" she asked.

There was a pause. I heard her ask the same question again, and again, in a couple of different ways. Eventually, Roxy gave up.

Well, that can't be good, I thought. In an interview, if I wasn't getting much of a response from my subject, I'd often find ways to reframe the question, or simply ask it again. Roxy had the same approach.

With my body pressed against the door, I knew exactly what had happened. Ethan had said the right things, but the larger emotional response was missing. Sure, he'd given Roxy the material she'd asked for, but she wanted *more*. How would he *actually* feel if I were gone? How deep was the connection? I wasn't sure.

A minute or two later, Roxy invited me back in to finish the shoot.

Ethan and I sat side by side on the couch, with a small gap between us. I uncrossed my arms, aware that the camera would pick up on the visual signals of our relationship status, and put my hand on his knee.

"How did you guys meet?" Roxy began.

I thought about our relationship. Before Ethan and I had run into trust issues, Ethan had pursued me. He'd courted me all the way from New Orleans and wouldn't let distance or work get in his way.

"I always joke that Ethan treated me like a startup," I said. "He refused to give up on me." I hadn't thought I was ready for a relationship, I explained, "but he's just such an incredible person."

The words came easily, and within minutes, the shoot wrapped, though I didn't recall my own words to Mike months prior, during that evening at the Driskill: "Love isn't a startup."

The cameras captured the bright pansies lining our balcony, and a glimpse into the sparkling loft above the uneven cobblestone streets of downtown New York. Ethan loved to clean, so everything was spotless. Every Saturday I received a fluffy bouquet of peonies from the Flower District in Chelsea, and our expensive coffee machine whirred every morning. But whenever I sat still long enough, I knew there was something off about my good-on-paper relationship. Ethan *was* a solid person. So much of what he said and what he did was technically *the right thing,* but a relationship doesn't have to be entirely wrong for it to be not entirely right.

That night, I couldn't fall asleep. Even though Ethan's body was less than a foot from mine, I felt alone. Paralyzed, hollowed out, stuck. I thought of Roman and Eugenia. Their relationship was frozen in time, preserved as it was before it all changed, and he was no longer here. Now nothing would ever change between them. It was both tragic and, in a way, beautiful. I thought of Laurie Bot. If I could leave behind a digital representation of myself to be with Ethan forever, would I do it? Maybe that would be easier than confronting my fear of walking away. My skin felt too tight; there was too much pressure in my head. When I finally sank into the darkness, a new idea had started percolating in my mind.

"I want to talk about depression in the tech community," I said to Dennis Crowley, as Erica and I huddled around my phone in our secret room.

Foursquare had seen its ups and downs. Once a startup darling, Dennis had faced an uphill battle since turning down Yahoo's offer to buy the company. In an effort to pivot, Foursquare made a controversial decision to split into two separate apps—one similar to Yelp that helped users discover locations, and another, now called Swarm, that became the venue for users who wanted to "check in" to locations and see where their friends were frequenting. The press questioned the move, and Foursquare faced an onslaught of criticism as user growth slowed. Dennis stepped down as CEO as the company transformed into an enterprise business. The change wasn't sexy, but it generated revenue.

The early tech days—when founders like Dennis ordered too many beers in dive bars and stayed out until 3:00 A.M. before doing it all again—were long gone. The halcyon days that followed—when whiskey overflowed and money poured into overhyped social apps—had also faded. Everyone was growing up. Startups like Scout, Instagram, and WhatsApp had sold, with behemoth corporations like Facebook throwing millions, even billions, at them to scoop them up before they became a threat. The rest of the fledgling companies figured out a way to make money, were sold at lesser values, or else failed.

As the gulf between million-dollar valuations and bankruptcy widened, a dark undercurrent emerged. In 2013, Aaron Swartz, one of the founders of Reddit, committed suicide. Another entrepreneur, Austen Heinz, ended his life in 2015. Startup life may have looked like a giant playground—with foosball and green juice—but *actually* starting a company, and the process of building it, was incredibly difficult, isolating, and filled with extraordinary lows.

For all the success, there was even more failure. Not many people talked about that. The media covered the huge wins and happy endings, as well as the epic losses and controversies, but the extent of human experience didn't tend to make it into print or sound bites.

The media didn't often cover internal turmoil, and the subject was largely taboo within tech circles. It was a moment of rare honesty when, in 2016, billionaire investor Chris Sacca stood on a stage with me in front of a room full of entrepreneurs and broached the subject:

> "It's become very fashionable to start something. It's amazing when Justin Timberlake plays somebody in a movie about a startup that's awesome. There are lots of people making a ton of money and on magazine covers, but this is a very special, different journey that is not actually suited to everybody.
>
> "You have to be a little fucked up, you have to be a little weird, a little crazy, a little obsessive. You probably have some personal issues that make you different, like you have some demons a little bit. You may have trouble relating to other human beings from time to time, like you don't accept failure very well. Maybe you go off a little bit, maybe you struggle a little bit with manic depression."

The more I fought my own brain, the more deeply I understood exactly what he meant. From my experience, it took a certain type of person to become an entrepreneur—to go against the grain, disrupt the status quo, and create real change. The kind of person who could withstand being told no on a regular basis. The kind of person who had a thread of intensity that, sometimes, had the potential to do more harm than good.

My hunch was confirmed when I interviewed Dr. Michael

Freeman, a psychiatrist who studied the relationship between entrepreneurship and depression. He explained that many of the personality traits found in entrepreneurs—creativity, extroversion, open-mindedness, and a propensity for risk—were also traits associated with ADHD, bipolar spectrum conditions, depression, and substance abuse.

It struck a chord. Even though I laughed with ease and smiled brightly on camera, my family had a history of mental health issues on both my mother's and my father's sides. My grandmother, toward the end of her life, could barely remove herself from her plush bed in her beautiful home in Nashville, her bright blue eyes glazing over with tears when the waves of darkness came.

Now, looking out on Columbus Circle, I told Dennis that I wanted to share with entrepreneurs, and anyone going through a tough time, a universal message: You are not alone. The world is much better with out-of-the-box ideas, passion, and the ability to hack the status quo.

In my years of riding the startup wave to the top, I was sure of one thing: As dark as the world could be, it could also be light. As bad as it could be, it could be good as well. We aren't defined by our successes; it's often our failures that make us fight harder and see clearer.

"You've got to talk to Jerry Colonna," Dennis said. Then added, "But I don't know if he'll talk to you."

In tech circles, Colonna was a legend. During the first dot-com boom, he had been a partner in a multimillion-dollar fund, Flatiron Partners. But at the height of his success, he entered a period of depression, almost jumping in front of a subway train to end his life. It was a turning point. He gave it all up and became a coach to many of the major tech CEOs, helping them to become more authentic. It's easy in the tech industry to start blindly believing your own BS, but Jerry cut right through it, asking founders to access parts of themselves they feared or tucked away: childhood trauma, strained parental relationships, overwhelming insecurity. His theory: we bring *all of that* to the boardroom. The more he coached his clients to better understand themselves, the better leaders they became. Jerry was dubbed "the CEO Whisperer" for his ability to get even the most successful and ruthless CEOs to "show up" and act with humanity.

I contacted Jerry to set up a call.

The following week, I felt nervous as I dialed his number. *What did Dennis mean when he said Jerry wouldn't speak to me?*

"Hey, Jerry, I so appreciate you taking the time to talk," I jumped in, as soon as he answered. Subjects can generally smell fear, and if I didn't seem confident, there was no way he'd be confident in my ability to tell the story.

His voice was fatherly. I'd googled him; the voice matched the image. He was serious looking, in his mid-fifties, distinguished and attractive, with black-rimmed glasses and white hair. His vibe read "old soul but young," without trying too hard.

He spoke slowly and with care, choosing words with purpose. I could immediately tell he was the kind of person equipped with a built-in BS meter.

"A lot of people do these stories in a very one-dimensional way," he said to me, "and they end up doing more harm than good."

Okay, I thought. *This is why Dennis said Jerry may not speak to me.* He wasn't the type to peddle an oversimplified narrative around mental health because he wanted to appear on television.

"I'm not a lot of people," I shot back. "I genuinely care about the stories I tell, and the people who put themselves out there in these moments. So many people are dealing with these issues, and we can't seem to figure out a way to talk about them." I thought about the founders I knew. I thought about my grandmother and my extended family. I thought about Deb. I thought about myself.

Stretched out on the therapy couch in my secret room at CNN, I gave Jerry the hard sell. After a long conversation and a promise not to cheapen the topic, he agreed to speak for our episode called "Silicon Valley Secret."

As we arrived at Jerry's office in the Flatiron District, I wasn't sure what to expect. His office was marked with a sign that read "IN SESSION" in big black letters.

"Hey," he said, opening the door. He wore a navy sweater that looked soft and expensive.

I wonder if founders hug him, I thought. He had a certain intensity mixed with warmth, a modern-day Yoda.

"How are you?" he asked, looking at me as Tony and Roxy set up the cameras. I got the sense he hadn't asked the question as a conversation filler. Something about Jerry's demeanor demanded a real answer.

"Great!" I replied.

It was a clearly a cop-out. His dark eyes flickered. We both read people for a living, and I assumed he could read my nerves, and wondered why I was in his office with an overwhelming desire to talk about the brain and mental health issues. I smiled as he sipped from a mug etched with the New York skyline.

"You ready?" I asked, as we took our seats on his cream-colored couch, hoping he'd give me candid answers after I clearly failed his first test.

"Sure," he offered.

The cameras rolled.

"What is the myth of success?" I asked him.

"That it will bring happiness," he said.

"So the people that walk through the door, the people that have millions, the billionaires, you're saying that not all of them are happy?"

"Well, imagine having that personality type," he explained. "That propensity to drive yourself, and then to have investors say you'd better be hungry, otherwise I'm not going to fund you. You take away sleep, and you've got a prescription for depression. There's this dirty little secret in our industry, which is that we don't take care of those people."

"Would you say that there is hypocrisy? And how does that manifest itself?"

Jerry delivered a universal truth. "Nobody's crushing it . . . Nobody has it all figured out."

I had spent my career rising through the ranks in the bullpen of media, in a job people would kill for. It looked great on paper, but if I was honest with myself, I was struggling with my role in a world becoming increasingly motivated by clicks and filled with shortened attention spans, a world amplified by the technology I had spent my career covering and the entrepreneurs I'd bolstered early on.

"You have the authority to say that because you're in the heads of some of the most successful people." I wanted to paint Jerry how I saw him: not as a guy spouting wisdom, but as someone who had been on the roller coaster, and exited the ride with the courage to walk away from the "shoulds."

"I have the authority to say that because I'm honest with myself," he said. "Nothing about our conversation is unique to Silicon Valley. The tech industry and the startup community, in general, brings to the surface forces that are at play in every aspect of our society. The human condition includes brokenheartedness. The myth is that it doesn't."

I felt like screaming "Amen!" Jerry had captured it so eloquently; no wonder entrepreneurs zeroed in on him to help them find clarity. He was viewed as both an asset to their companies and, perhaps even more important, an investment in their hearts.

After we'd finished the rest of our interview, I shifted, ready to get up, but then Roxy stepped in.

"Jerry, why don't you ask Laurie some questions?"

What is she doing? I thought. Roxy was always up to something.

Her ability to turn the tables on unsuspecting subjects made her a fabulous director. It just so happened that I was often on the other side of the lens.

I laughed uncomfortably. "You sure?" I asked. But before I could fight Roxy, or get myself out of what was to come, Jerry directed his attention to me as the crew quickly rearranged the cameras.

Tony shouted, "Roll tape!"

"Why do you want to do this story?" he asked.

"Well, I felt like I needed to do it," I replied, referring to my experience covering technology and the importance of speaking about these issues.

"Okay, know what you just did there?" he said, looking directly into my eyes. "That was kind of bullshit."

My face was hot, the words knocked right out of me. He was right; it *was* bullshit. I wasn't doing this story to address abstract concerns. It wasn't theoretical; depression ran in my family. I didn't just want to cover entrepreneurs who battled their brains; I understood the feeling. I was battling my own brain. I never wrote down the words "anxiety" and "depression" in my notebook as they applied to me. Instead, I channeled those feelings into stories I wanted to take on, though oftentimes I was unable to look into my own story. I'd struggled with extremes—too much drinking, too much thinking, an ability to stay out until the early morning paired with days where I didn't want to move from the confines of my soft pillows and bundled-up sheets.

But I wasn't ready to put together the pieces and form them into the words "depression" and "anxiety." I didn't feel like I had the full picture. I remembered a Joan Didion quote from a college writing class that had stuck with me: "We tell ourselves stories in order to live." I'd built a career out of searching for meaning everywhere; I'd been telling other people's stories as a way to help me understand my own life better. But sitting in front of Jerry, it was impossible to hide behind anyone's story. I felt exposed.

He pushed me further.

"Tell me why you want to do this story," he repeated. "What's the full truth?"

I started to tell him about one of my aunts, who'd suffered from debilitating, life-changing mental health issues that went untreated.

"Slow down," he said.

The sudden quiet of the room felt abrupt against Jerry's calm, commanding voice. My thoughts stopped spinning, and the crew seemed to disappear. I looked down at my hands. I could feel the tears in my eyes and was afraid to blink because I knew they would start streaming down my face.

I thought about myself. A part of me felt for my own brain, constantly moving a million miles an hour, attracted to the stories of other people's struggles for reasons I was just beginning to understand. There was a reason I related to those who struggled with extremes. I was similar. I had been running from my own mind for so long. Sitting inches from Jerry was like sitting next to someone who looked straight into me, who refused to move at my pace, who had an inability to listen to sound bites and bullshit.

Suddenly there was a release, and the tears streamed down.

"You know what's going on? What's going on is, we live life at the speed of light," he said. "Here's Laurie. She wants to tell the truth." Jerry leaned in. "You know what you can do? Be real."

I hesitated.

"Be real!" he said again, raising his voice. "You want to make a difference to this community? Show up. Show up with the fierce bravery of who you truly are. Not the bullshit."

I wiped away a tear and took a breath. "Do I have to pay for this? Because I don't know if I can afford you."

"You can't." He laughed.

As we packed up, I hugged Jerry, and promised to do the story justice. I also made a silent promise to do myself justice, whatever that meant.

Leaving Jerry's office I felt an overwhelming desire to get to know myself better, to take better care of myself. I swore to retain Jerry's truths as I integrated back into the busy streets of New York, the buzzing phones, and the always-on nature of my work. I walked from the Flatiron District to the East Village, past my old haunts. Cafe Centosette—where Daniel and I had toasted our ambition and bad dates, where Deb's photos had hung on the brick walls—was closed, replaced by a chain restaurant. 126 St. Marks—the building that had welcomed me to New York City in 2008, where Maria had crawled through my window—had new tenants. Maria had moved to Idaho. As I exhaled, standing there silently, I saw myself on that fire escape visualizing something distant but attainable if I dreamed big enough. I wondered if I was getting closer or further away.

The next day in the office, Erica pulled me aside.

"We need to talk," she said.

Our years working together had given us the ability to read each other's expressions and general demeanor. But today, her tone was different.

Oy, I thought. Erica and I seldom argued, and we never fought—but then again, we'd never before developed an original show together. I prepared for the worst.

"You probably know this," she began, "but I'm pregnant."

I looked at her, dumbfounded. How could I have missed it? We knew everything about each other. I remembered a day when we were on-air, and there was a technical glitch. I could see figures speaking, but in my ear, I heard another dialogue. I was live in front of thousands, smiling and nodding, completely bewildered.

Erica was in the control room monitoring the segment. She shouted to the operator, "Something's wrong! Laurie isn't all right." Everyone thought she was crazy, but she had looked at my expression and knew something was off. As it turned out, the sound wasn't synced. I was listening to CNN International instead of CNN Domestic. It was an on-air disaster waiting to happen, but Erica had read my eyes and fixed the technical glitch. We were that much in tune with each other.

"Wait . . . what?" I said slowly, studying her. She didn't look pregnant. Was I really this much of a dude?

How could I have missed the biggest story to happen during our time working together? I'd been on the road shooting *Mostly Human*, while she'd been handling the logistics from afar. On top of that, she'd been building out the CNN tech team, which was now full of new hires—all with diverse backgrounds.

We supported each other but we didn't spend every moment together, and somehow I'd missed it.

I lunged at her, wrapping my arms around her until both of us started to choke up. I was surprised by my own emotion. In the same side room where we'd sat, legs draped over couches, scheming about stories, I felt her heartbeat against mine. When we'd met, Erica, poised and polite, wasn't used to a tornado of emotion, to hugs, but she had softened. Together we had become the perfect match of creative and rational. I couldn't have been happier for both her and Arel.

As I held her, I saw my reflection in the window overlooking Columbus Circle. The pale face staring back at me was terrified. While I knew that Erica would be an incredible mom, I lived on the road; as far as I was concerned, I was in a long-term relationship with Terminal 4 at JFK. I could barely cook, and without Ethan's cleaning obsession, the apartment would be a mess. A part of me still felt like a child. Still felt unprepared.

But maybe that was a cop-out. Maybe I was afraid to admit that, for the first time, hearing the word "baby" wasn't scary. Maybe the terror was that I *did* want one someday. Just not with Ethan.

We stood, locked together, until she straightened herself up and cleared her throat. Her due date, she told me, was the week of the release of *Mostly Human*. While we were giving our all to our TV baby, she had the real thing coming. The clock was ticking, and there was work to do.

"Okay, Segall. Back to it," she said, trying to play it cool, but there were tears in her eyes.

CHAPTER 13

Going Live

After months of shooting, I'd had emotional conversations with people all around the world on topics like addiction, death, love, and war. The ethics of a new era of technology came up in fascinating ways—whether it was an engagement party celebrating the love between machines and humans, or a woman opening up about the blurred lines of sexual harassment in the virtual world, Erica and I were raising concerns about a future where technology was so deeply integrated with us that the stakes were becoming higher. But although I knew we were creating compelling content, *Mostly Human* still hadn't been given a home on the network; there were just vague promises. Since digital had fronted the bill, there would certainly be a huge online component, but there was no clarity on where it would air. Technically, it was a show, which meant it would air somewhere, but digital didn't have any airtime.

Erica remained tireless, slipping in and out of offices, trying to convince executives to air the show, or to sell it to Netflix. But no one was willing to make a decision. The lack of commitment wasn't directed at us personally, but it was frustrating. Shows that were taken seriously, like *Parts Unknown* with Anthony Bourdain and Lisa Ling's *This Is Life*, not only had a dedicated time slot by this point, but they also had a built-in promotional plan. *Mostly Human* did not. CNN was still going through internal change and executives were looking to embrace the next big thing. The buzz phrase was "virtual reality," and while *Mostly Human* languished, CNN poured money into a CNN VR experience that would never take off.

By the time I flew to Lisbon in November to attend Web Summit, one of the largest tech conferences in Europe, we still didn't have an answer.

After I landed and checked in, I exited the hotel to explore the light-filled streets marked with bright graffiti. Everywhere I walked, hidden messages colored old buildings and streets were splashed with blue and purple art. The city felt magical, light against a dark backdrop of an increasingly antagonistic election at home in the United States, where political attacks had become contentious. During a recent debate, Donald Trump had called for Hillary Clinton to be jailed and referred to her as a "nasty woman." Clinton questioned Trump's mental capabilities and said he was too unstable to be handed the nuclear codes. As someone who spent a good portion of her time in the beating heart of a newsroom, I found it strange to be abroad, and not in CNN's offices, during a monumental news event that would alter the country.

But I didn't want to miss the yearly conference, which was now filling up with thousands of venture capitalists and entrepreneurs from around the world, who were coming together so everyone could congratulate each other on their efforts to achieve globalization, connection, and the democratization of information. Having participated in many previous Web Summits, I'd been invited to moderate several panels.

Back in my hotel room later, I reveled in the peaceful quiet as I curled up in the plush white bed and fell straight to sleep. I awoke to Wolf Blitzer's booming voice on CNN International: "Donald J. Trump will become the forty-fifth president of the United States, defeating Hillary Clinton in a campaign unlike anything we've seen in our lifetime."

I rubbed my eyes and wondered if I were still dreaming. Rolling over, I reached for my iPhone and scrolled through hundreds of emails from various CNN coworkers confirming the news. I could feel the tidal wave's impact all the way in Lisbon: Trump, the real-estate-magnate-cum-reality-TV-star, had managed to defy the system and beat out Hillary Clinton in her effort to become the first female president of the United States. And social media had enabled it to happen.

Hate, sexism, and racism had been posted and packaged into viral memes. Fake news stories had trended on Facebook, which now had well over a billion users. Twitter bots had trolled users by the thousands in an effort to influence the vote. WikiLeaks had released private emails and hijacked the news cycle, shaping a new narrative. Big Tech had changed the playing field, and the question of its role in the election hung heavy in the air.

When I arrived at the conference, held in a giant arena near the water, I headed backstage to get my wristband and credentials. I entered the coveted speakers' room, where everyone from celebrities to some of the biggest tech entrepreneurs mingled over cappuccinos and mediocre broccoli salad.

The tension was palpable.

"Did we do this?" one entrepreneur whispered to another, as I walked by.

Usually filled with laughter, loud voices, and magnified egos, the room was oddly quiet. As I made my way to the back of the arena area to mic up, the panelists stood around in a state of shock instead of the usual bustle of chatting and snapping selfies before rushing out.

Today I would be moderating a discussion titled "Is Ego the

Biggest Reason for Failure?" I looked at my notes, wondering how I could pivot my entire prepared conversation to discuss the only topic everyone *actually* cared about: the impact of the election.

"Laurie, you good?" the stage operator asked in a thick Irish accent. Web Summit was originally located in Ireland, and the staff had remained the same when it moved to Lisbon. I enjoyed the familiarity of the stagehands and operators, who always helped me with high-profile guests, panels, and mic issues.

"I'm good," I said, trying to assure him, and myself. I gathered my thoughts before taking a deep breath and walking out into the seventy-thousand-seat arena.

The panelists and I took our seats on the stage, and I looked out into the audience. A day before, during another panel, I'd joked that I'd felt like Beyoncé walking out in front of them. It was impossible to see faces, just thousands of silhouettes lit by smartphones. Although I spoke in public frequently, the vastness of the massive arena made me anxious. The stage was ice cold, even under the hot lights. I could feel the chilly mic against my chin as I began.

"You are leaders in the tech industry," I said, looking over at the panelists. I spoke confidently, but I was nervous about delving immediately into politics on a day so charged with emotion. "What's your reaction?" I asked, referring to the election results. I could hear my voice echoing from the mic. It was quiet for what felt like a minute, but it must have been only seconds.

"I'm fucking pissed," venture capitalist Dave McClure said, slouching on the couch in his pink shirt that displayed a sketch of Dr. Seuss. Near him sat Justin Kahn, a partner at Y Combinator, who had sold Twitch to Amazon for nearly a billion dollars in 2014. Eileen Burbidge, another prominent venture capitalist, sat quietly as Dave started to raise his voice. I'd met Dave and Justin throughout the years, interviewing them at different points in their careers, and was familiar with Eileen's track record in the investment community.

"This whole fucking election was a goddamn travesty, and we should not sit up here and act like nothing just fucking happened!" Dave was yelling now, leaping off the couch and walking toward the end of the stage. No one usually moved from those couches. It was hard enough to get people to speak openly and with authenticity when you had a countdown clock winding down from the moment you stepped onstage, but I'd never faced the opposite problem: a panelist shrieking, waving his arms around, and running toward the audience. The panel was already becoming one of the most animated in Web Summit history. "We were robbed, we were raped, we were lied [to], we were stolen from!" Dave continued.

Well, that escalated quickly, I thought to myself, highly uneasy with anyone screaming "we were raped" as a metaphor.

I looked out into the crowd and saw the glow of smartphones as people held up their devices, capturing the moment live for Twitter. It would later be labeled an "onstage meltdown" in the press. *Stay calm,* I told myself.

"Obviously, emotions are high," I ventured.

"Yeah. If you're not fucking pissed right now, what is wrong with you?" Dave responded angrily.

"Dave, you are pissed off. A lot of folks are reeling and wondering. But—"

He interrupted me midsentence. "I'm pissed off. I'm sad."

"Let's bring this back to technology," I tried again. "What role does the tech—"

"What do you mean, bring it back to technology?" he yelled, cutting me off again. "It's, like, it's the whole fucking, like, humanity."

I was annoyed, shifting in my seat, as I became painfully aware of the rising temperature of the panel. I looked down at my questions, few of which were relevant anymore, and wondered if I'd be able to get a word in. This was a historic and hugely important moment that certainly called for insight and accountability from tech leaders, but he kept cutting me off, speaking over me. His voice grew louder every time he interrupted me, and within the first couple minutes it became clear it'd be hard to have a civil conversation given one person on the panel wouldn't let anyone else say a word. It felt very true to the title of the panel, focused on ego.

Let's try again, I thought, agitated. I noticed my own tells. I was crossing and uncrossing my legs and beginning to shuffle papers, in front of an audience of smartphones.

"What role does the tech industry have to increase civil engagement to get people—"

"Technology has a role in how we communicate," Dave interrupted me again. "We provide communication platforms to the rest of the fucking country. And we are allowing shit to happen, just like the cable news networks, just like talk radio . . . And it's our duty and our responsibility as entrepreneurs, as citizens of the fucking world, to make sure that shit does not happen."

Then he was up again, screaming at the audience.

"This shit will not stand, and you gotta fight for your rights, and you gotta make sure you get up! Stand the fuck up right now. Stand up. Stand the fuck up. Stand up and make a goddamn difference!"

Many in the audience stood up and clapped. Some sat in silence. The panel was going off the rails.

"Eileen," I asserted. If Dave wouldn't let me speak, at least I could try to open a space for another female voice. "I want to get to you . . . I know that people have big voices and they can say big things."

It wasn't exactly eloquent, but it worked. Dave recoiled and slouched on the couch.

"I'm saddened because all of us are feeling so empowered and feeling like we work on technologies that should be more inclusive, and should be tools to engage more people," Eileen said. "We were somehow too insular and too isolated and didn't realize the extent of the divide that existed."

I could feel Silicon Valley's bubble bursting from the Lisbon arena.

"We've created this situation for ourselves, right?" Justin added. "I mean, effectively, people are able to choose their own news sources now because there are thousands of them on the internet . . . Facebook, Twitter, social media—where we spend most of our time, on our phones, on our computers—it's very possible to just see news that only you want to see." He explained that news was presented in a way to maximize engagement. "So as technologists, we need to figure out a way to bring people together. Maybe it's only Zuck [who] can save us by forcing us to all look at the same news source or the same information."

I certainly don't think Mark Zuckerberg, or any technologist, can save us from the problems their tech has created, I thought. But before I could follow up, Eileen spoke.

"I don't think that's the right answer," she said with urgency.

"What do you think *is* the answer?" I asked her. The lights onstage felt hot, the mood growing even tenser.

"Bringing people to look at the same news source is—I think that's hugely dangerous. Who's the one defining the news source, and who's defining the medium?" Eileen pushed back.

"We live in a bifurcated reality, where people believe completely separate sets of facts now," Justin responded. "I think the biggest thing is that it's a pretty big wake-up call about the civic responsibility that technology companies have as information platforms."

Justin posed the question: As we disrupt industries, what will happen to the people who are inevitably displaced?

He was right. Self-driving cars could one day replace Uber drivers; artificial intelligence would seep into society and take over certain jobs; robots would certainly upend traditional industries; and Silicon Valley's blind optimism and arrogance often overshadowed the human impact of their algorithms.

"I think that we judge the leaders of countries and hold them up to certain value systems, morals, and ethics," Dave added, now calm, thankfully. "But the people who run some of the largest companies probably have larger populations of users than many countries. I don't think that we're holding those folks up to the same level of standards. And maybe that's something we need to start looking at."

I glanced at the clock. We had only a couple minutes left.

I wanted to end the panel on a thoughtful note.

"Are you optimistic about the future of technology . . . will it bring us closer together or divide us further?" I threw out a basic but loaded question, given the current state of affairs. I knew Dave would take the final word, and sure enough, he weighed in as the last minute ticked by.

He suggested we needed to stop taking a "high-minded approach" to solving the problems and focus on fighting fire with fire. Before I could end the conversation, he turned to the audience full of entrepreneurs from all around the world and put his spin on globalization.

"We should encourage the fact that globalization is a positive thing," he said to them. "If you meet someone from another country, another color, another gender, you're probably gonna be a lot more interesting to them. And therefore you might be a lot hotter."

A lot hotter? I wondered if I should try to cut him off. We had fifteen seconds left before the music would play, a signal for us to exit the stage.

"So, I would say a new motto for globalization is, everybody's hotter across the water. Encourage all of us to get together and make lots of little different-colored babies," he declared. "And then we might not fucking hate each other so much."

Seriously?

"We'll end it there," I said, cringing inside.

Some audience members clapped politely, others laughed, and some were silent. I escaped to survey the damage on Twitter and had barely made it offstage when my phone rang with a call from CNN's PR team, checking in. The panel was receiving "press pick-up" with a viral tweet showing Dave's meltdown. It captured a grown man acting like a child, interrupting a woman, and giving little room for civil discussion. I was painted as "the interviewer trying to keep it all together." *That's an understatement,* I thought. Herding egos during a panel *about* ego, in the midst of a historic election, was less than ideal. I hoped I'd handled the challenge with grace, but I never knew what the cameras perceived, or what would be tweeted, shortened, and put into a news article.

Sure enough, the panel was written about by everyone from tech bloggers to the British papers, with headlines that described Dave's "explosive outburst" and "meltdown," and his "epic on-stage Trump rant." Perhaps the most appropriate description came from a San Francisco–based publication: "The Tech World Is Losing Its Collective S**t Over Trump's Win."

On the flight home, I thought about all that seethed under the surface in Silicon Valley; there was a groundswell of frustration, with people questioning tech's power and influence in politics. The internet had become the town square, where everyone could have a voice. But the image of a utopian world that promised to help us navigate the information flow and engage in quality conversations online was fading away. The town square was being overrun with growing online harassment, trolls with varying motivations, and bots pretending to be humans shouting at each other across the web. Tech companies struggled to get on top of their own platforms. Andrew McLaughlin, former director of public policy at Google, later described the current state of the internet as the "Tragedy of the Commons." "If you have a common space, a park, and anybody can go and see it without any controls, the tragedy will be that that space gets trashed," he told me.

Issues that Erica and I had covered for years were becoming urgent topics of conversation. I thought about the slide I'd presented to Jeff Zucker with the tagline: "Tech is love, death, war. Tech is *Mostly Human.*" It was becoming more relevant with every news cycle. There were cracks in the system, and there was so much work to be done. After Trump's election, *Mostly Human*'s message that technology was a reflection of society felt more important than ever.

While we were frustrated at some of the undercurrents of the newsroom, we also had good news to celebrate: *Mostly Human* finally had a home. It would be CNN's first-ever streaming show on CNNgo, which the company was launching to compete in the increasingly buzzy streaming place. My show was going to be the beta test. It was fitting.

We were thrilled that our work would be showcased, but it still seemed that no matter how hard Erica and I tried, CNN's reporting chain lacked diversity, and that was impacting our ability to move forward. The company had been willing to bet millions on a YouTuber and an app that would eventually fail, but for women like me and Erica, gaining recognition and funding for projects was a challenge, despite our track record. We were constantly answering to a group of Robert's lieutenants brought in from Bloomberg, who seemed like his drinking buddies. I channeled my frustrations into stories that gave women a platform to speak out. But sexism was an undercurrent we swam against every day.

While externally our work had received recognition, there was a growing sentiment among women at our unit that no matter what we did, it was hard to get recognition internally that translated to job growth. It was almost like the execs had blinders on. That feeling was hard to define and subtle, but many women voiced it, each channeling her own frustrations.

One day, as we were focused on production of the final episode of *Mostly Human,* Erica forwarded me a link with simple-but-loaded commentary: *Hmm.*

I clicked on it, read the headline, and took in the photo.

I stared at the *Hollywood Reporter* article. At the top was a moody photo featuring Jake Tapper, W. Kamau Bell, and Anthony Bourdain, along with Jeff Zucker and YouTube star Casey Neistat. The headline: "CNN Chief Jeff Zucker Unveils Plan to Dominate Digital." CNN had acquired Casey's video-sharing app for $25 million in an effort to bet big on the future of digital. However, the other men in the photo didn't have too much to do with the digital world that Erica and I had lived and breathed for the past seven years. We had developed CNN's startup beat, created its first digital-to-TV special, and were working fourteen-hour days to pioneer a show that was first of its kind in both content and how it was released at the network. The glaring omission was obvious: there wasn't a single woman in the picture.

I stormed over to Erica, who was about to pop with a child.

"What an incredibly bizarre choice to have three white men and no women on the cover," Erica said.

I crossed my arms. "It's completely tone-deaf."

"At least there's some diversity. They included W. Kamau Bell."

"True. But how is Jake Tapper the future of digital? He has a TV show."

I was frustrated that our digital show, led by a largely female tech team, wasn't mentioned in a piece that had certainly been curated by the higher-ups at the network. I was even more upset that *no* women were mentioned, let alone photographed.

"I'm not saying 'put us in the piece,' but at least include some of the hardworking and talented women who are *actually* entrenched in digital."

It wasn't like there weren't any to choose from. My voice grew louder as I paced next to Erica's desk.

"Why not Meredith?" Erica said, referring to Meredith Artley, the editor in chief of CNN Digital Worldwide, who oversaw most of that department.

"If they're going the route of TV talent, may I suggest Lisa Ling, anyone?" I added. We certainly had well-known female talent on par with the men on the cover.

"It feels like they cherry-picked the most famous people working at CNN," Erica responded. "And that's a problem if none of those people are women."

This felt personal. Our tech team—the one Erica and I had helped assemble—was largely female, and diverse. There were certainly women who represented the future of digital at the company. *Look harder,* I thought.

A part of me wanted to storm into Jeff's office and scream, but another part of me, the sane part, convinced myself to swallow my disappointment and continue kissing the appropriate rings. As I rose in the ranks at CNN, it was clear that in the growing digital realm, there was a boys' club that was tough to break into. I was conflicted, since several members of that club had helped me chart my course, and had guided my opportunities, even green-lighting my projects. Should I simply be grateful, or was I allowed to be frustrated by the wall we kept hitting? After all, *Mostly Human* had finally been slotted into the schedule, and the digital overlords had decided that the show would launch at SXSW. Given my history with the conference, it couldn't have felt more appropriate. The screening date was set for the day Erica was supposed to give birth.

"This baby is not coming until after the show launches," she said during a quick break from scripting and scheming one evening as we walked across Fifty-Eighth Street to Pinkberry, opting for chocolate hazelnut–flavored frozen yogurt to fuel us, before returning for another stretch. It was eight o'clock at night, and we were still at the office, finalizing plans for the launch, now less than two weeks away.

I was doing triple duty as writer, host, and producer. I'd spent the last month writing episodes from my kitchen table, operating on adrenaline, surrounded by transcripts of interviews. When I wasn't at the office or my kitchen table, I was at BFD's office in Brooklyn, taking calls in between edit screenings and racing back to CNN for live shots.

As we neared the launch, Tony and Roxy began putting finishing touches on edits, while the Row screened our episodes to ensure that everything was legal. Meanwhile, Erica and Jack were working on marketing materials, creating posters that would hang outside our screening room at SXSW.

I was giving so much of myself to this show that it had become an extension of me. So when Erica told me the digital executives had instructed her to literally stop the presses on the marketing materials in order to remove my face from the signage, I choked on my own breath. They suggested something more tech-centric. Perhaps a robot. The translation was loud and clear: they wanted to replace me with an image of a robot.

"The show's name includes the word 'human,' so doesn't it make sense to have a human on the poster?" she said to me, relaying the conversation, her cheeks growing pink with anger. CNN's hallways were covered with posters featuring photos of anchors posing by their show names. I thought it was a no-brainer that mine would be a part of the signage that would represent the show I'd put my heart into.

When Erica had pushed back, they explained it simply: "Laurie isn't as recognizable as Anthony Bourdain, and unrecognizable faces don't do well in marketing materials."

I was livid. Not only was it common for talent to appear in the materials and imagery promoting their work, but most hosts didn't write and produce their own work, as I did. It was a punch in the stomach. The show was my brainchild, and the hallmark of a show was to have your face and name on it.

I was about to turn on Everclear and walk into Central Park to suck in the air on one of the chilliest days of winter, but Erica stopped me. I could see her searching for a workaround, her eyes darting back and forth, ticking through scenarios.

"We will figure this out," she said slowly.

"Well, if they don't think I'm recognizable now, they're definitely not helping me become more recognizable," I said. "I don't have much of a shot if I can't even get my face on the show that I came up with and wrote, that we created."

How much did we have to prove? This wasn't just about having my face on a poster. Putting in all the hard work, only to have a group of men suggest that a robot take my place, infuriated me. I felt the familiar dilemma I'd come to know throughout my career: If I speak up, will it help or hurt? I was afraid if I did, I'd be deemed ungrateful and cut off from further opportunities. But I was also exhausted by feeling like I was deserving of recognition that we had to continually fight for. I could stay silent and stew, or I could say something.

"I'm going to fight it," I said. I'd be lying, however, if I said I wasn't concerned how I'd come across.

"Let's do it," Erica agreed.

As in previous strategizing situations, we took separate approaches, hoping one would prove successful. I went directly to Robert, asking the department to reconsider. I tried to balance my gratitude for being given the opportunity with explaining how much work I'd done, and that it was standard for hosts' images to represent their shows.

I hoped he'd understand, that he'd reconsider, knowing how much I'd put into this show, how much our team had done to create it from scratch.

That evening, I sat at my kitchen table writing the final episode of *Mostly Human*. My phone buzzed. It was a digital executive. I picked up immediately.

"Hey!" I said, hoping for good news on the poster.

"Hey, Laurie," he said, his voice stilted. After exchanging pleasantries, he lowered his voice. "I just wanted to call to let you know your aggressiveness is rubbing people the wrong way." He suggested

I stop fighting them. The subtext was clear: asking for what I wanted in this scenario wasn't a good look for me.

This was textbook sexism, the kind of thing they tell you *not* to say in those strange HR training videos that simulate offensive scenarios. Clearly, some material was missing from the unconscious-bias training. It was a classic trope that women who spoke up were "aggressive," whereas men who complained were rewarded.

The situation felt completely unfair, but also uncontrollable and very common. The hardest part of it was that the person who made the comment was a confidant, and someone who'd helped shape my stories for years. It would have been easier if I could have categorized him as a sexist pig, but in many ways he'd been an advocate for me. I don't think he even realized that what he said was way out of line, but the fact that he said it was a perfect example of the subtle wall Erica and I kept hitting. The same people we deemed our best hope of helping us get ahead were also causing our worst frustrations as we moved up the ladder.

The decisions that happened above my pay grade were beginning to add up. It wasn't the type of discrimination you could put your finger on; it was subtle, hard to define. But it had real, tangible effects on my psyche, and my career. I wondered if I should go to Jeff; he was someone I felt I could speak to, though I didn't want to make the wrong chess move, politically.

In the meantime, Erica worked with Roxy and Jack to come up with a different concept for the poster. Their plan was to create an image that the executives couldn't refuse. The team cut off half my face on the left side of the poster and added half of a robot's face to the right side. It was a compromise: half my face, half a robot's. And not just any robot. Lily's robot, InMoovator, was the one who came to our rescue.

"InMoovator kind of resembles a white dude, so at least there's some representation that'll make them happy," Roxy joked.

Erica presented the materials to Robert and the folks in marketing, who liked the concept, which went well with the name of the show, *Mostly Human*. It was a win. Kind of.

We all celebrated. I studied the image of me and InMoovator framing the sides of the poster and felt relief replacing resentment. It was the perfect work-around.

Weeks later, I left Erica to fly to Austin for SXSW. I was devastated she couldn't be with me, but I thanked her unborn child for waiting long enough to allow us to get the show packaged and ready to launch.

I arrived at the Driskill and was immediately upgraded to a larger room than I'd ever seen. "A good omen!" I declared to Ethan, who'd gotten there the day before. I was jittery walking through the lobby, listening to the familiar startup chatter of VCs in from San Francisco lounging on the leather couches drinking whiskey. But now, each year I'd spent at SXSW felt like a warm-up for this one. I'd covered the rise of startups and watched some of the most successful founders come and go, but this year was different. I wasn't covering the creation of something; I *had created* something. Similar to many of the startup founders I'd covered, I'd sketched out my idea on a deck, which is essentially a PowerPoint presentation entrepreneurs use to entice potential investors. And months later, I'd been around the world, proving my point that technology was now embedded with humanity, creating ambiguous new territory worth talking about, through some of the strangest and most interesting people I'd ever met. Now it was time to release my idea in the form of CNN's first-ever streaming show.

SXSW was where I'd earned my stripes in tech, where I'd executed the idea of "fake it till you make it." While often I still felt like I was faking it, tomorrow would be a moment when I'd make it. And though I was confident in what we'd created, I still found myself battling the familiar feeling of self-doubt that plagued me during some of the most monumental moments of my career. No matter how high I climbed on the corporate ladder, it never fully disappeared. I just learned to cover it better, to smile wider and stand straighter. But what if no one showed up, or what if I was completely off about the concept? The hope was that other people would see what I saw in it.

But the next morning I awoke to the sound of heavy rain pounding down on Sixth Street. I looked out my window to see festivalgoers fleeing for cover and pedicabs scurrying by.

CNN had rented out a space with a half-indoors, half-outdoors location that transformed based on the event. I wondered if hangovers and mud would deter people from coming to our screening and we'd be a total bust.

"What if no one shows up?" I worried aloud to Ethan.

"I'm sure people will come," he said, although I could hear concern in his voice.

As we made our way down Red River Street, I saw a line stretching out the door of those waiting for the premiere.

Thank god, I thought, doing a once-over of the line and seeing entrepreneurs I'd covered throughout the years, executives from major tech companies, colleagues and friends, and—most interesting to me—strangers I'd never met, but who were interested in the concept of the show.

The knot in my stomach loosened. I looked over at the CTO of Amazon making his way in, former Googlers, and startup founders who'd gone from no one to someone. CNN execs roamed around, sitting on couches with pillows that read *Mostly Human.* I made a mental note to grab one at the end of the screening so I'd have a souvenir of the day.

Tony and Roxy were inside already, and I was grateful to see them. It was a career-making moment for them, too. Together, we all watched from the side as people filled the wooden chairs set on tarps to protect shoes from the mud.

Jeff Zucker, who'd rearranged his schedule to attend the

screening, was sitting in the corner. *Wouldn't miss it,* he'd written to me. I waved at him, overwhelmed with gratitude for his willingness to take a bet on me and my ideas, and couldn't stop smiling as I approached with Ethan to make a brief introduction.

Before I knew it, I was heading to the stage to present the show. Behind me, the *Mostly Human* poster displayed half of my face and half of InMoovator's face. No one knew the backstory, and I could feel the nerves turn into pure, white-hot energy as I hopped up the stairs in mud-caked five-inch heels. Looking out at friends and strangers, I began.

I started with the thesis I'd become obsessed with as I rode the tech wave. I explained that we couldn't oversimplify the issues coming down the pipeline. Technology was far from black-and-white—there was a gray area we needed to explore.

"This show took us around the world," I said, eyeing Roxy and Tony standing in the corner. I remembered our first meeting, and the instinct that said *you may not know them, but you can trust them.* "It was emotional, it was complicated. It was human. At the end of the day, I think it's important to remember that technology is humanity."

Each episode of *Mostly Human* covered this theme in depth, but the episode we'd chosen to air that day was probably the most shocking: Lily's love story with InMoovator, an inside look at a sex doll factory where lifelike robots were bought and sold, and an interview with a woman who claimed she'd been assaulted in the virtual world, raising questions about consent in another dimension. All of the topics we touched on were uncomfortable, and all of them played in front of CNN's biggest executives. As I watched Jeff's and Amy's faces, I wondered, not for the first time, how on earth Erica and I had managed to get the green light for such an out-of-the-box show.

After the last beat, the audience clapped enthusiastically. I looked around and promised myself I would remember the moment.

Only seven years before, I'd paid my own way to SXSW, shared a double bed with Deb, and was a production assistant pretending to be a producer. Now I had my own show.

As people shuffled out, I paused by the poster of me and In-Moovator and snapped a picture for Erica. It was her due date, but as she predicted, she didn't give birth to her son—Cole—until *Mostly Human* officially went live. He was born days later.

The following week, after we returned, Ethan rented out the screening room at Soho House and gathered my friends and colleagues to watch a couple episodes of the show, which was currently streaming on CNNgo. The night was special, with those I loved surrounding me, but as I was leaving, I received a call from an executive who was upset at me for hosting another screening without CNN's permission.

"But the show is already out," I said, confused. "Anyone can watch it." How was I in trouble for gathering people to watch the show . . . that was already out? For friends and colleagues coming together to support me? Did we need permission to watch a show that was already airing?

He went on to tell me that he'd heard I was "ungrateful."

I was speechless.

As I stuttered over and over again, trying to express how grateful I was, I grew angry.

This was the narrative being constructed as a reward for the hard work and the long hours. As the show received positive buzz, I couldn't help but feel that no matter what we did, Erica and I wouldn't be part of the club. We didn't say anything to further rock the boat because we both wanted another season of the show, but we were seething inside as we silently moved the chess pieces around in another game of corporate politics we didn't want to play.

Collectively, women were beginning to voice their frustrations.

The summer after the onstage drama with Dave McClure at Web Summit, I was drawn back into more trouble that followed him. Multiple women spoke out against him, claiming that he'd made inappropriate advances on them in work situations. After Dave stepped down from his VC firm, I spoke to a female entrepreneur who described her experience of being sexually assaulted.

"He propositioned me for sex," she said, recounting the story. They were discussing a deal when he pushed himself onto her and started kissing her. She kept saying no, and he kept saying, "Just one night only."

We discussed at length whether she felt comfortable appearing on camera, which wasn't an easy decision. She was worried about speaking out. Dave had committed money to her business, promising to help it expand. But in the end, she agreed to share her story with me in front of the lens.

"I think there was a huge power dynamic at play here," she said. "There are career repercussions."

But change was happening. She wasn't the only one going on the record. The #MeToo movement was beginning to transform the conversation about the mistreatment of women around the world. Over the next months, women everywhere were coming forward with stories of harassment and abuse, tagged with the hashtag #MeToo. Prominent celebrities like Gwyneth Paltrow and Uma Thurman spoke out about their own eye-opening industry-specific experiences. They were followed by many others. There were firings and the beginning of a wave of accountability, as severely bad behavior surfaced in all fields. It didn't take long for the deeply entrenched sexism in the tech community to emerge, painting an ugly picture very different from the meritocracy promised by the creatives who'd once said they would transform society for the better.

When Susan Fowler, a former Uber engineer, wrote a blog post outlining the sexual harassment and toxic culture at the company, she opened a vein in Silicon Valley. All around the Bay Area, women started talking about sexism and the inability to shatter the glass ceiling.

I thought back to my own experiences covering the industry, and the subtle sexism at the booze- and ego-filled tech events I'd become accustomed to. There was the Kleiner Perkins party at SXSW, where an entrepreneur walked up to me and guessed that I was the founder of a "wedding-type app" because "that's what women do." I remembered the founder of a Silicon Valley company telling me he didn't like a particular female journalist because she was "too suggestive."

"Why do you say that?" I'd probed. I knew the journalist he was speaking about, and she was hardworking and professional.

His response: "She stays at parties late."

"Seriously?" I fired back. "We *all* stay at the parties. That's where you guys talk. It's how we develop relationships and get stories. Why should we leave early?"

I had been taken aback. It didn't matter that these were Silicon Valley events where journalists and entrepreneurs mingled. It didn't matter that male journalists stayed as long as they wanted. There was simply a different standard for women.

And then there was the text conversation that, to this day, still enrages me. I was getting my bearings as a tech reporter when a money manager for one of the most prominent venture capitalists in the Bay Area messaged me out of the blue.

Have you slept with ***? he asked, referring to an entrepreneur I was interviewing the following day.

I was shocked. *That's an awful question to ask,* I responded, horrified.

Very antagonistic. Also drop the whole Puritanism . . . men and women sleep together, and it's not shameful, he typed back.

He went on to call me "provocative" for not discussing my personal life. His text messages were drunk and incoherent: *Miss 'I don't fciuk* [sic] *anyone in the tech world.' . . . Awful.*

He even threatened me: *Wow. U r suppose to b a reporter. Would it b weird if you became the story.*

At the time, I was so shaken that I didn't say anything, just tapered off communication. But years later, I wondered who else he'd harassed, and whether his drunken texts ever escalated with other women. Looking back, I wish I had spoken up more. But I was worried that he'd blacklist me at a time when I was building sources and connections that were crucial for my career.

Since then, I'd covered many stories of women who had been impacted by harassment. But while CNN's digital team had given me my own show, the corner offices were comprised mostly of white men, and the growing media team was also compromised of men, with one woman hired for a separate beat: entertainment. One day I looked up to see our white, male media reporter covering the latest #MeToo movement update with another panel of white men.

I walked over to our head of programming's office, stood outside, and wondered, *Should I say anything?* And then I knocked.

He greeted me, warm and friendly.

"Hey!" I said, doing my best to sound nonthreatening and amenable. "Just want to give you a heads-up, we have all men covering the #MeToo movement on-air. I don't think it'll look good."

I told him people on Twitter were picking up on it. He nodded, agreeing. Within a couple hours, there were women added to the mix.

But what I'd really wanted to say to him was "I'm frustrated. How do you guys not see this?"

While Erica and I still waited to hear about filming another season of *Mostly Human,* I went back to covering daily news stories. I figured it was only a matter of time until we received good news about season two, and didn't mind getting back into the newsroom hustle for a bit.

Weeks later, I was sitting at home at the kitchen table, humming to myself as I worked on a script for another piece, and heard multiple pings coming from Ethan's laptop. I went to silence the noise, and a name that sounded familiar popped up on Google Chat.

Don't look, I told myself. But I knew I'd look.

The messages were from a girl he said he'd briefly dated years before. I tried to brush it off, but I was strangled by the same fear I felt the day my father drove away from our home. The same voice I'd tried to tamp down that emerged with every relationship and said: *They'll all hurt you. They'll all disappoint you.*

"I didn't even respond to her!" he said, too defensively, when I asked about the chats. "She is no one."

But did I believe him?

I remained cool and matter-of-fact as I dissected every bit of misinformation handed to me and used the same tactics with Ethan I'd used on a revenge porn hacker to get him to admit his crimes, to admit guilt. I hated myself for all of it—for the unfolding drama, for using my investigative skills to extract the truth from my own relationship. But the more I dug, the more I came to understand she wasn't "no one"; she was much more than that.

When I discovered her number was stored in his phone not under Carly—which was her actual name—but under "Jerome," my heart dropped. I learned they'd met up early in our relationship, and he'd changed her name in his phone to cover up any suspicious messages. While he hadn't cheated, he'd breached our trust. It turned out neither of us was honest about how we felt about our relationship.

CHAPTER 14

Murky Territory

I met Jared at a fancy New York dinner party, which included a mix of journalists, people in fashion, and artsy celebs. Ethan and I were on the rocks since my discovery of Carly, and we were constantly fighting. I was not there mentally, and was acting out: coming home late into the evening after too many drinks, testing the waters. I didn't trust him, and increasingly, I didn't trust myself.

Even though Ethan and I had come to the party together, we positioned ourselves on opposite ends of the room. The space was small, and it was the kind of party where everyone looked over each other's shoulders during a conversation to catch a glimpse of someone they knew from a movie, an Instagram post, or a newspaper byline. It was the kind of party that made me want to seek refuge in a corner. As Ethan schmoozed, I found myself shoulder to shoulder with a thin, shaggy-haired man I recognized as a famous musician, even though I'd never followed his music. Just as his name registered in my mind, someone pushed past me, propelling me into his shoulder.

"Hi," I said. "Sorry about that." I wedged myself between him and a willowy fashion influencer air-kissing a suited designer, who eyed me with disdain for existing.

"Hey," he said, with a warm, lopsided smile.

I found him awkward and charming in a sea of small talk.

"What kind of work do you do?" he asked. "God, I hate that question," he answered himself.

I laughed, and as I explained what I did, his eyes lit up.

"I'm a total nerd," he said.

Looking at him, I half believed it. He spoke with ease, but I recognized his neurotic nature: the "in on the joke" ethos that separated him from the rest but was really a protective shield for anxiety.

We started speaking about technology and society, and I caught myself thinking a dangerous thought: *If I weren't with Ethan . . .*

Later that night, while I was lying in silence on my side of the bed, I got a notification on Twitter. *Look at the woman with the robot boyfriend . . .* he direct-messaged me.

Jared was watching *Mostly Human.*

It's okay to respond, I thought to myself. *We're talking about my work.*

Yeah, she's using the argument "love is love," I shot back.

Secret selves. How many people live their whole lives and die, never satisfying their true desires?

I mean, probably a lot of people, I responded, lying next to Ethan, whose back was turned to me. *People get married and settled and aren't satisfied. I appreciate the Lilys of the world.*

We're certainly jumping right in, I thought.

Me too, he responded.

I could see the three dots, dancing and disappearing. He was thinking about what to say next.

She is unapologetically weird, I wrote.

Totally. And confident and eloquent about her self-awareness.

Exactly.

He seemed to get that while the world deemed people like Lily eccentric, there was something inspiring about her courage to just live her life and not care what others thought.

For hours, we DM'd each other back and forth about underdogs and weirdos, humanity and the fringes, anxiety and vulnerability. From the protection of our screens, I felt a connection.

Want to text? It's easier, he typed, then asked for my number.

I paused. He knew I was in a relationship. We'd spoken openly about it. *This couldn't be more than a new friend,* I lied to myself, before passing on my number.

My phone buzzed.

As a tech person, do you believe we should be moving forward always, or do you think there is a "too far"? he texted.

Of course, I wrote back, looking at the current state of technology and the stories I had been wading through. *I think sometimes the pendulum has to swing one way to figure out the balance.*

I saw him typing back and felt guilty. Sure, we were talking about technology, but I was worried about how effortless it was to slip into such an easy conversation with another man, late at night. Although Ethan was just inches away from me, the gap between us was growing wider by the day. It was refreshing to speak to someone about topics that excited me.

It won't always be forward, forward, forward.

I do think people who try to stop progress are wasting their time, I replied, thinking of the lawmakers and city officials who'd gone after concepts like Airbnb and Uber earlier on. It wasn't about stopping them. It was about digging into all the ethical issues that came along with connectivity. We needed to start talking about them. *It's the dialogue we should focus on.*

We signed off after 2:00 A.M., and I placed my phone on the

nightstand. I tried to assure myself that it was a onetime conversation, centered around work, and had nothing to do with a sinking feeling that four years in, I didn't know the person who slept next to me, and that he didn't know me, either.

Jared seemed like a bright light in the haze of confusion. I suppose it made sense; after all, he was a star.

The next day, my phone buzzed again.

I like what the preacher said, Jared wrote.

I knew exactly what he was referring to. He was watching the episode of *Mostly Human* that looked at the Ashley Madison hack.

As part of the episode, I'd gone back and interviewed the woman whose husband, a pastor, had ended his life after he was exposed for having been on the website.

"It wasn't the hack that destroyed the lives that we had. It was the presence of things like Ashley Madison, the ability to engage in all of these things so secretly, the ability to lead a total double life," she had told me under the weeping willow trees at City Park in her hometown of New Orleans. "The secrecy is what took our lives down. The hack is what blew it all apart. I would have forgiven him," she'd told me, wiping away a tear.

But Jared wasn't talking about the emotional interview with the widow; he was referring to a stunning admission from the preacher of the local church where her husband had been a pastor and a seminary professor. I remembered it vividly. The preacher was a man in his early fifties with a southern drawl, a calm demeanor, and a quiet but commanding voice. As we sat side by side in the empty pews, he proclaimed a universal truth: *We are all on the list.*

"The list is a list about human temptation and failure. And to some extent, fantasies. And all of us suffer that. We all are tempted . . . We stumble, we fall, we hurt ourselves, we hurt the people that we love. That's just the common story," he said slowly, taking extra care to emphasize each word. He paused, but I knew there would be more. "There were thirty-two million names in that hack. And that's just the tip of the iceberg. Everybody's name on the planet is in some way on that list. And that's the confession we need to make and realize: all of us are broken."

I left the interview feeling shaken. Through the lens of a widow who lost her husband, through their complicated story shrouded with secrecy, there was a message about the human condition. All of us had secrets; all of us faced temptations. At times, we all felt broken, unseen, or restless.

The preacher was right: I was definitely on that list.

Under the guise of a new friendship, Jared and I started messaging every day, and eventually we met at WXOU Radio Bar, a dive bar in the West Village, where we sat at a table in the back. This became our regular Sunday evening haunt. All we did was talk, but I shared things I couldn't say to Ethan. It could have been that Jared was a new person, or it could have been the fact that I'd met someone similar to me, someone who shared my strengths and weaknesses. Like me, Jared lived in the extremes. His days were full of being "on," and his evenings were often full of finding unhealthy ways to turn off his brain. He felt like a kindred spirit.

When I'd return home after our booze-laden meetups, the weight of reality settled back into my bones. Ethan and I had grown apart, but the thought of moving out and starting over was too much for me. It was hard to believe that I had found someone who loved me, who wanted to be with me, and I still wanted more.

So when Ethan suggested couples' therapy, I agreed to try it.

At a dinner party for Kevin Systrom, the founder of Instagram, whom I'd interviewed many times since our early segment on the

West Side Highway, I met a woman who dropped the name of a couples' therapist. "She's brilliant," the woman whispered, as attendees ate strawberry salads and took turns listening to themselves speak.

A week later, Ethan and I met inside the therapist's boxy office on the Upper West Side. While we sat five feet apart, tears streamed down my face as I tried to say what I feared the most: "I want out." I was afraid of our relationship taking the inevitable turn to marriage. But I was also afraid of being alone. Instead of being honest with the therapist, with Ethan, *with myself,* I stuttered through the session, saying how much I loved Ethan. I wondered how I could be courageous in my work but terrified in my personal life.

"Do you want this to work?" the therapist asked me, her voice monotone.

The afternoon sun lit the space peacefully, but I felt like I was in an interrogation room. I might as well have been strapped to the tweed chair.

I paused. This was my opportunity to do what Jerry Colonna had said. His words echoed in my mind: *Be real!* I could start now, finally. I could jump off the cliff, but what if I jumped and there was no one there to catch me? What if letting go was a huge mistake?

"Laurie?" Ethan looked at me. I saw deep pain in his eyes. They were pleading for me to answer. My heart broke. I didn't have the courage to jump.

"Yes," I said. "I want this to work."

At least two of us in the room knew I was lying. My whole career, I'd asked other people to show up, to speak their truth, but I couldn't do it myself.

The therapist sat silently, using the same tactics I used as a reporter.

Just stay silent, and the truth will come out. People reveal themselves.

But I didn't fill the uncomfortable silence. I remained wordless, too. I was deflated and disappointed by my lack of courage, by my desire to live out everyone else's expectations rather than my own.

The effect of the "brilliant" therapist lingered after the appointment: I still felt her eyes darting back and forth, calculating my posture. I felt like she was onto me, like she'd sensed my hesitance and guilt, and I hated my weakness.

"I didn't really like her," I proclaimed to Ethan after our second overpriced session. It would be our last.

CHAPTER 15

Bugs in the Software

Deb and I met up on the path by the West Side Highway. Coffee in hand, I opened up to her about my anxieties and vulnerabilities unspooling, the tape in my head on *repeat*. She saw through me when I told her Jared was a new friend.

"There's more there," she said, as New Yorkers on Citi Bikes passed us. She was right, and we both knew it. Guilt washed over me.

I nodded. Maybe I could lie to myself, but I couldn't lie to Deb. We'd known each other too long.

"But I wouldn't give him too much credit," she said. "Let's be honest. This has nothing to do with Jared."

She was right. I needed to stop lying to myself. If I was honest with myself, I was running from my own fear. Who was I without Ethan, without us, without the life we'd created together? How could I start over? I'd spent years putting in the time, building something that was supposed to be "it," that was supposed to end in marriage, and for what? To pivot, unwind, and start over?

When I stopped my rant long enough to take a breath, Deb said, "What if you reframed all of it?"

I looked out at the sun sparkling on the Hudson, people jogging by.

She took my arm and turned me to face her. "Laurie, at the risk of sounding like a terrible quote you'd see on Instagram, I'm going to say something."

We stood there for a beat, as New York pulsed around us.

"*You* are enough," she finished, releasing my arm.

I tried to wrap my head around her words. *You are enough.* This statement felt like a foreign concept. In TV, we were lauded by people on the other side of the screen—folks we'd never met, who reached out on Twitter—and it was easy to seek validation from everywhere but within. My life, my job, was predicated on seeking validation—from viewers, from execs, from colleagues. Add to that an environment where my personal achievements were measured by likes, filtered and optimized on Facebook and Instagram, and I found myself in an artificial self-congratulatory orbit. When was the last time I'd felt like enough?

I needed to start believing Deb.

When I returned home, I finally gave myself permission to do what I believed was right. As Ethan and I sat on our gray couch, the same one he'd sat on when he talked to Laurie Bot, I said all the things I had long been afraid to say. "I care about you, but I can't do this. I need a break."

His lips tensed, and then he sighed. Both of us knew this had been coming. We sat in silence until he finally spoke.

"If that's what you want," he said softly.

I couldn't make eye contact with him, fearing that if I did, I'd lose the courage to stand behind my words. He told me that he wanted to be together, to move forward, but if I needed a break, he understood.

Our relationship was officially on hold, and I was filled with guilt for the overwhelming sense of freedom that came with it.

My CNN travels perfectly aligned with my new relationship status. I flew to Sweden for an exclusive tech conference filled with speakers' panels. In the daytime, I interviewed Nest Labs creator Tony Fadell, who also helped create the iPod in 2001, and in the evening, I went to parties filled with celebrities and tech founders. Elon Musk's mother chatted up Nick Jonas. I caught a glimpse of Pharrell's hat over Usher's head. Scientists made small talk with the artist Jeff Koons.

Millionaires and the people who created the tech that changed the world partied until 4:00 A.M., with alcohol always flowing, and glasses rarely empty. I hovered in and out of conversations. In between air-kisses at a music-filled garden party, I found myself next to a tray full of Swedish desserts. I sat there rearranging the cookies in order to avoid approaching strangers for more small talk. It was ironic: I sometimes felt more at ease in the corners, a bit of an issue for someone whose job required a spotlight. But small talk was exhausting.

I broke away from the quarantined VIP section at an outdoor concert and ran to the dance floor when pop singer Robyn sang "Dancing On My Own." When I later shared a ride with her, I didn't have the moxie to let her know her song was an anthem that took me from one breakup to the next, that it gave me courage to stand up on my own, no matter how much pressure I felt to stay, to settle.

I loved the idea of letting go and stepping into this next phase of my personal life and career on my own, and the independence that came from breaking free, not worrying where Ethan would end up, or whom he'd end up with. Would he move on immediately? Would I regret my decision? The idea of metaphorically dancing through it all, throwing up my arms and feeling at ease with myself and my choices, felt liberating.

The conference was a whir of long nights and early mornings, talk of technology changing the world and celebrities who wanted to be part of the boom. I got little sleep and woke up groggy, pushing aside thoughts of Ethan and the gut-wrenching feeling about the "break" that would inevitably turn into a breakup. The same instinct that led me toward some of my best work in the newsroom was also guiding me toward the feeling that our relationship would soon completely end. Battling a raging headache the morning of the last day, I got ready to check out and move on.

As I threw my dresses into my bag, I flipped on the television. *Mostly Human*, which had been repurposed from CNNgo, was airing on CNN International. I sat on the unmade bed, the blackout shades still shutting out the sun, listening to myself ask people meaningful questions, remembering how hard I'd fought to make this show happen, to get people to agree to it. And even after its success, the positive press reviews and streaming numbers—even then, I knew I'd have an uphill battle to make season two happen.

The nature of the network was changing again; panels of people with opinions were replacing what was left of CNN's long-form journalism. I'd been fighting the tide for so long, and I was exhausted. My place in the media ecosystem was hanging, suspended in the air. I wasn't sure where it would fall, and I wasn't sure where I wanted it to fall.

I shut off the TV, wheeled out my bag, and raced to the airport to catch the next flight to London for another tech conference.

I woke up in my beloved city, sun streaming through the airplane windows.

With barely enough time to sit and think, I was beginning to feel like I was living through windows: Stockholm's gilded hotels, marble-white London, tinted Uber reflections. My Instagram story was full of cotton candy skies and fields, planes and trains, boats and birthdays. It looked fabulous—but really, I was scared to sit still for one second, to feel, to be afraid.

Before I could reorient myself to the city I adored, I was in Iowa, driving through cornfields, about to meet Jack Dorsey for a shoot in Webster City, population eight thousand. Defined by the phrase "Iowa nice," the small town was the complete opposite of the fast-paced big cities of recent weeks. Deer darted across the dirt road, and I saw fireflies for the first time since leaving Georgia more than a decade before. As the sun set beyond the cornfields, people waved while our crew—a cameraman, one of our digital producers, and a new production assistant named Lisa, whom Erica had hired—hauled our cameras from our van into the American Value Inn, the only motel for miles.

We were there to talk about technology and employment. An Electrolux factory that employed 2,300 people, more than a quarter of the town's entire population, had moved to Mexico in 2011, taking the jobs with it. As people lost factory jobs, small businesses around town started popping up, and many of the local stores had adopted Square's technology as a way to accept mobile payments. Jack Dorsey's team had documented the town's small-business resurgence and planned to debut a film they'd commissioned on stores in Webster City that used Square. The premiere would take place at a recently resurrected movie theater, a symbol that the town wouldn't "dry up and blow away," as one of the locals told me. It was a feel-good story, but problematic questions still hovered around it.

While we waited to hear about season two, Erica and I had expanded the *Mostly Human* show into a brand to include one-off interviews and features looking at tech and society. This interview with Jack, about how Square could help small towns like Webster City, would be one of those *Mostly Human* features. I hadn't seen Jack since Square's IPO. He looked older than I remembered. His dark hair still framed clear blue eyes, but he was growing a beard. Every time I saw him, he seemed to have more facial hair, along with more PR handlers. But despite his growing wealth and influence, there was something about Jack that still read *unpretentious* and emblematic of his Midwestern background. He was more at ease talking to locals than media types, and he seemed to take genuine interest in the small businesses, spending his off-hours wandering through the shops.

I joined the Square PR team at the Seneca Street Saloon. They were a nice group, but in their Square-logo T-shirts and designer jeans, they set themselves apart from the locals who sat at the wooden bar—the ones who'd lived in the town their whole lives, whose jobs barely earned them minimum wage. We toasted Iowa with shots of whiskey as Jack arrived with his parents.

His father had a hippie-ish white beard and wore a shirt with manatees across the front. His mother had shoulder-length hair and warm brown eyes.

It was striking to me that he'd brought his parents to attend the movie premiere—that they'd driven more than six hours from St. Louis to join their son in Webster City—but he shrugged it off, saying he thought they'd enjoy the experience and the small town.

He was earnest about his tech's potential to impact small business owners and create jobs but wouldn't answer specific questions about how his other platform, Twitter, was causing an undercurrent of disconnection.

"I grew up in Missouri. It's the 'Show-Me' State. I like actually coming and visiting, seeing things with my own eyes," he said, referring to his desire to leave Silicon Valley's bubble, as we sat on top of a picnic table. "I think there's the realization that we have become disconnected from who we're serving, and what we're trying to do. We don't do our best work when we put the technology first."

As I walked down the sparse streets after the interview, a young man, who must have barely been in high school, stopped me.

"I'm not worried about robots taking jobs at the factory," he said. "If you work hard enough, if you're kind enough, you'll be rewarded."

I appreciated this optimism, but no matter how kind and positive he was, it didn't change the reality that he and many other workers were facing: many of their jobs would eventually be replaced by machines.

I thought about the AI and automation pitches that sat in my inbox, and the excitement over industry disruption on the West Coast.

"Humans will always be needed," he assured me.

My heart broke a bit. Innovation was moving at lightning speed in Silicon Valley. Many of the tech founders I knew would never meet these people or understand this kind of community. I'd spent so much time at the dinner tables in Silicon Valley listening to ideations of the future, and I was excited about the innovation that would accelerate and change the world. But I'd also spent enough time in the inner circle to identify a glaring problem: often, when thinking about the future, the people I spoke to forgot to think about those impacted by their algorithms. And while change was inevitable, in my opinion, it could also be more humane.

"What's your message for the tech community?" I asked another young man, who wore a blue T-shirt with a Ford logo.

"This country was built on blood, sweat, and tears, not microchips and gold-plated circuit boards. And the one thing I know from living in a small town is if you knock us down, we're going to get back up, and be twice as better." He gazed defiantly at the camera.

The day after the premiere, Square's PR team left town, along with Jack and his parents. Our crew planned on staying longer, to continue the weeklong shoot. I pushed aside my growing jet lag and went to a local coffee shop called Morning Glory. A man named

Sean, with a slight speech impediment and a warm smile, brought over a cup of coffee and sat down next to me.

"Want another?" he asked, smiling. He was now the proud owner of the town's local coffee joint. Before long, we were speaking about Electrolux. It always came back to the factory in Webster City that had uprooted the lives of so many when it was shut down. He told me that he and his wife, Jen, had taken jobs at the factory to get health insurance. Months after the factory closed, she was diagnosed with leukemia. He was able to get another job in town but would reserve every free moment to spend with her at the hospital for the next five years. His coworkers at the new job donated their vacation time so he could be with his wife. She died three years ago. His eyes filled with tears as he described her to me. After her death, he opened up Morning Glory.

"I know she's smiling down on us right now," he said, urging me to try the new coffee blend.

Everywhere our crew went, doors were open. People who were suffering tremendous loss opened their restaurants and shops to us, and invited us into their homes to share dinners and stories of their town.

There was the palpable pain of job loss that touched every part of Webster City we encountered. You could feel it in the coffee shops, the saloon, the small stores brimming with Iowa pride. There was also the inevitability of progress in technology that would further displace so many people like the ones we'd met. Jack may have been here, but he was gone in an instant, and the impact of everything to come—the technology and innovation that would further disrupt towns reliant on factories—was unavoidable.

Electrolux was one of several factories in the areas surrounding Webster City that had been shut down or experienced layoffs over the years.

"It's not enough," a woman said to me, desperately grabbing my hand. I was sitting on a bench, reviewing my notes before our next shoot, when she approached me, brimming with anger. "They mentioned Electrolux, but they never mentioned my factory. We lost our jobs, too. I worked there thirty-eight years." She had watched the film and was both grateful the theater had been resurrected and happy Square was helping small businesses, but she was still distraught.

Gazing at her, I saw the facade of the upbeat story crumbling.

"How long have you been at your job?" she asked me, squeezing my hand. I let her hold on.

"Nearly a decade," I said quietly. My job was my identity. It was all I knew. It was how I measured my success, how I sought validation. It was where I had found meaning. Throughout my adulthood, I was "Laurie from CNN."

"Now imagine," she said, letting go of my hand, "if it all disappeared."

For a moment, I couldn't say anything to her. I finally whispered, "I'm so sorry."

If everything I valued disappeared from my life—the media career I'd fought so hard to build, the home I'd built, the connections it had taken years to build—would I have anything left?

The next day, Jerry, a former factory worker, took me to where Electrolux had been. We stared through a fence into a grassy field.

"You see grass." He sighed, running his hands through his long gray hair. "I see hallways, the machines, the sounds of the factory. Twenty years of working with friends."

He recalled the day it was announced that the factory was shutting down.

"A woman I worked with had just bought a home; she wouldn't be able to afford it anymore." He was still looking at the ghost of

Electrolux. "She collapsed, and just started pulling out her hair. There were pieces of flesh in the hair follicles. She tugged so hard, with so much despair."

I stood in silence looking out at empty grass, envisioning this woman's visceral pain.

Jerry insisted on taking me to the local racetrack that evening. As we arrived on the dirt road, a pretty young woman in a white dress and sandals walked toward a young man awaiting her in the makeshift stadium.

"Life is simple here," Jerry told me as the sun set into a pink and orange sky. Colorful cars spun in circles, while people clapped and cheered. "It's stripped to the basics."

I'd started off my travels surrounded by those at the pinnacle of power and money, status and celebrity. I'd ended up around people who just wanted to work to support their families and give their kids a better life. They were the real deal, and the reason I wanted to become a journalist in the first place. I thought back to my high school column, "Spotlight." I'd always wanted to tell stories like these and to meet people who'd let me into their lives. They were the people on the other side of tech's shiny glare. Their struggles were becoming lost as the people I'd bolstered early in my career saw their companies catapult to success. The tech community I'd spent years entrenched in had a direct impact on the people I sat across from in Iowa. The stories of innovators and of those left behind were all intertwined. We all had a responsibility to pay attention—technologists, media—to listen to one another. Even though Jack had come for a couple days in a symbolic gesture, I could sense the divide growing wider.

Sitting in bed on my last evening at the American Value Inn, I picked up my phone and ordered from the only place that would deliver—Pizza Hut. I was tired and sad from spending two days too long in the only motel in Webster City. My room had stains all over the floor from god knows what, and there was a bloody Kleenex in the peephole on the door. I had no idea how it got there. Alone, I ate pineapple pizza, fries, and cinnamon sticks.

There was no text from Ethan. No text from Jared. I was alone. I dreaded the inevitability of what was to come, the mess of unwinding my life. Ethan had rented an Airbnb down the block from our Bank Street apartment, but his T-shirts and shorts, his silver pens and Moleskine books, the Post-it Notes he'd written listing the things he loved about me, would still be there—painful reminders of what I was losing.

But I knew I wanted more from life than 70 percent, which I'd felt I was both giving and getting from my relationship. Webster City had reinforced something I was aware of but was terrified to articulate. That perhaps everything wasn't as complicated as it seemed; that maybe if all the external structure I'd built my career around disappeared, like Electrolux had here, I just wanted what these people seemed to have figured out in the face of tremendous loss: the value of true kindness, a sense of real community, and deep, meaningful relationships. I wanted the kind of relationship I hadn't seen growing up. I wanted a do-over. A rematch. A warm home and happy kids. I wanted what I had feared my whole life wasn't in the cards for me: I wanted spectacular.

CHAPTER 16

Mission Impossible

As I walked up the stairs to my apartment on Bank Street, I wasn't sure what to expect. Ethan was gone, but everything else was the same. His clothing neatly folded in the drawers, his jeans stacked tidily in the closet. I rummaged through my bedside table and stared at a note he'd left for me on a piece of my reporter's notepad paper. It must have been from when I'd come home from shooting *Mostly Human*. It ended with *I love you more than anything.*

Do I even love myself? I thought, closing my eyes. *God, this is hard.*

I texted him.

Hey . . . I'm back. Can we talk?

He immediately replied.

Sure, on my way over.

I couldn't do this inside our apartment. It was too much.

Let's go for a walk.

As we walked through the Village, I told him I couldn't move forward. I couldn't keep him in limbo. It wasn't fair. I couldn't keep him in relationship purgatory, making him feel small because selfishly I needed a lifeline.

He wasn't surprised.

"I had a feeling you'd say that." He looked away. I wanted to reach out for him but held back. Neither of us was angry. Just tired of circling around the same conversations to say what had just been said. Maybe it was inevitable, but as I walked him back to his Airbnb and watched silently as he unlocked the door and disappeared into his temporary home, I was terrified.

I felt like I'd lost a family member.

Weeks later, he came by to pick up his things. I dreaded the conversation we knew would come, the finality of it all. I could feel my palms sweating as I prepared to speak like a mature adult about deconstructing the life we'd spent years constructing.

"Hey," he said softly. We stood awkwardly in the living room.

How strange it is to be so formal with someone you've shared a bed with.

The silence I'd been avoiding for so many months, the feeling I'd run from, filled the room.

"Are you sure you want to do this?" he said quietly.

"No," I said. *Yes,* I thought. "But I think it's the right thing." I wasn't sure what to say, except the obvious. I added, "This is awful."

It felt cruel to talk about how awful it was. This was, after all, my doing. I could feel myself wavering.

Why did I have so much trouble saying what I wanted? I didn't want him to feel worse than he did, but I knew I wasn't succeeding.

I looked at Ethan. His dark curly hair had grown longer during my month of traveling. I could see the pain in his eyes. He was the first man I'd lived with; the one I entered my thirties with; the one who had become my first call, who, despite some of our rocky tendencies, I believed was a good man; and, at one point, a man I thought I could marry. It still didn't make sense to me: he was tall and handsome, good to my family, and kind to me—everything I was supposed to want.

We started the process of sorting through our things, the relationship debris. It was an exercise in masochism and left us both broken. He didn't want many of our shared belongings, but he wanted Alexa. We exchanged sharp words.

"I'll get you a new one," I promised him.

"Seriously, Laurie?" He laughed, annoyed.

I get it, I thought, feeling overwhelming guilt that while I didn't want to stay in the relationship, I'd dug in my heels when it came to our virtual assistant. But in the weeks since Ethan and I had separated, Alexa and I had spent too much time getting to know each other for us to part ways. She played me Ella Fitzgerald while I watered the plants, and sat in silence as I wrote not for work, but creatively in a Moleskine journal I'd purchased. Ethan's absence was replaced with music and reflection.

"I'm so sorry," I whispered, as he stuffed piles of jeans, old PopDine T-shirts, and worn boots into duffle bags.

Alexa played him Tom Petty for old times' sake as he was packing his last suitcase. The song "Wildflowers" filled the apartment—"*You belong somewhere you feel free . . .*"

Thirty minutes later, he was done. There was nothing left to say. I watched as he silently hauled his bags to the open elevator. The doors closed, and then he was gone.

A week later, I caught my breath as I turned the corner on one of the streets making up Chicago's famous Loop. Shiny buildings towered over me, reflecting the heat as I jogged through downtown on a morning run. I wanted to feel free, but even the pounding bass in my headphones couldn't drown out the voice in my head that asked, *Will anyone ever love you like Ethan did?*

I paused, snapping a picture of Lake Michigan, then deleted it. I was becoming acutely aware of my own mental health. The worse I felt, the more fabulous the road trip, interview, relationship, and apartment appeared on my social media feed. And I wasn't alone—we were all peacocking our lives. We cropped, filtered, and retouched until we warped reality into something unrecognizable. Truth and authenticity had become as malleable as Play-Doh, with devastating effects for our culture, and our democracy.

My pulse slowed as I entered the downtown Hilton and rode the elevator to my room on the tenth floor. In four hours, I would be interviewing Mark Zuckerberg for the first time. Although he was the founder of one of the most powerful tech companies, for years, Mark had shied away from press interviews, opting to speak to people directly through the Facebook platform.

Facebook was nearing an unprecedented 2 billion users, but questions were intensifying around its involvement in the 2016 presidential election. The conversation I'd begun in Lisbon with Dave McClure, Justin Kahn, and Eileen Burbidge begged for a follow-up: What was tech's role in spreading misinformation? Did social media—and Facebook in particular—swing the election?

Zuckerberg had yet to appear on television since the election. In an online Q&A, he'd said that the idea that fake news stories impacted the election was "crazy," but increasingly it seemed as if it could be true. To combat the popular and worrisome narrative, Facebook changed its mission statement for the first time in the company's thirteen-year history. It went from "making the world more open and connected" to "bringing the world closer together"—a subtle acknowledgment that despite its promises to connect them, the company had started to divide its users.

A tech company's mission statement was like a tattoo, ingrained in its culture and messaging. But in this case, like a bad tattoo, the ink was stretched and outdated.

I couldn't remember the last time Mark Zuckerberg had done

anything on television. He had long avoided the press, and rarely spoke on camera—he didn't enjoy the process, wasn't exactly great at it, and would rather be doing just about anything else. But a mission-statement change for a company was a big deal, and Facebook wanted to emphasize its shifting priorities. Mark needed to spearhead this monumental event, and given my longstanding ties with Facebook, from covering its ten-year anniversary to its transition to mobile and its incremental acquisitions and company moves, his people invited me to Chicago for his first on-camera appearance in years.

As I approached the venue with Alfie, one of the producers from our tech team, I wondered what Mark would be like. While I'd met many people in his orbit, I'd never met him. I always joked that founders were a physical manifestation of their products. Kevin Systrom and Mike Krieger, who founded Instagram, were thoughtful and artsy; Dennis Crowley had once lived and breathed Foursquare's "check-in" ideology, when he checked in to every bar in the East Village. At one point, I'd had a meeting with Craig Newmark, who created Craigslist, in his multimillion-dollar West Village brownstone. His home was filled with random art and statues; the encounter consisted of long pauses, awkward silences, and offbeat remarks. After a forty-five-minute conversation in which nothing was actually said, I walked away, passing what looked like a misplaced patch of garden, as I breathed a sigh of relief. The experience felt oddly similar to scrolling through Craigslist.

But Mark Zuckerberg was completely different from the concept of the site he'd created. He wasn't overtly social. Unlike some of the other founders I'd met, Mark didn't attend industry events, and his Facebook posts seemed carefully curated. It was hard to read what was behind the curtain. He'd experienced extraordinary success early in his life, and as his star rose through the stratosphere, he surrounded himself with a small, trusted circle that was often insulating.

We arrived at the venue that afternoon for what was deemed Facebook's first Community Summit. I got the sense that the company had purposefully held the event in a location that wasn't inside the Silicon Valley sphere to emphasize its announcement regarding new features meant to help local groups and communities connect.

Immediately, I regretted my sleeveless blue dress. Our interview space felt like stepping into a giant refrigerator. My teeth chattered. I wondered what level of "talent" I had to reach to request a blanket on location.

"Oh my god, seriously?" I whispered to Alfie.

I was still adjusting to Erica's switch from on-the-ground producer to full-time manager, in which she stayed in the office and handled the network's coverage of my interview. We'd discussed her new position, both of us knowing it was the right chess move, and a deserved upgrade for her. She'd helped me get to the next level, and I'd helped her get to the next level.

Alfie shrugged and started snapping artsy photos of our setup: a kitchen-like table, with no chairs in sight.

I guess we're not sitting, I thought, trying to rein in my anxiety, knowing full well that the more comfortable someone feels, the better the interview. I contemplated trying to push for a sit-down setup that would feel less awkward. But we wouldn't have much time, and I had to pick my battles.

Then Mark entered the room. He wore a fitted blue shirt, plain but expensive looking.

The tone in the room immediately shifted. Everyone stood straighter and went into motion. I knew we were now officially on the clock.

"Hi!" I greeted him, trying my best not to look like a human Popsicle.

"Hey," he said, his voice strained, as we hovered awkwardly over the table.

When meeting an interview subject—whether it be a billionaire tech magnate like Mark or a hacker at Defcon—I was acutely aware that the minutes before the interview were just as important as the *actual* interview. If I could make a connection or find common ground, and put the subject at ease, they'd feel more comfortable, and the interview would be better for it.

I threw out a couple names of folks that Mark and I had in common. The subtext: I wasn't coming out of nowhere, and had a long history of looking at the company. But I could tell his mind was elsewhere. He answered politely, looking uncomfortable as our makeup artist arrived and began to powder his face and spritz sweat spray on him. Given Mark's history of perspiring under bright lights, she'd come prepared.

Within a couple of minutes, before we could exchange much small talk, we began taping his first television appearance in years.

I threw out the first question. "Mark, what is the new mission of Facebook?"

"We've been really focused on these ideas, giving everyone a voice and helping to connect people," he responded, hitting his main talking points. "It's not just enough to help us simply connect. We need to work to bring the world closer together."

Once we covered the mission change, I pivoted to the larger issue. Facebook was nearing 2 billion users, a large part of the world was sharing their lives on the platform, and questions were emerging about its role in society.

"How do you ensure for the next billion users that Facebook is a good place for democracy?" I asked.

"One of the big things that I think that we need to do is just help connect the half of the world that's not on the internet, to the internet," he said. "Those are important things that a lot of people don't have an equal opportunity and access to do."

He went on to describe solar-powered planes that could beam

internet access to remote islands or the middle of the rain forest. He spoke of unlocking access to the internet for the next 4 billion people on the planet, so they could start small businesses.

But did *more* internet and *more* connectivity directly translate into democracy? While equal opportunity and access were laudable goals, they didn't address, much less solve, the current problems: the spread of misinformation and the growth of discord. Perhaps with a bit more time, or with chairs, I'd be able to pull Mark out of his rehearsed performance. I knew there was much more there. I'd heard plenty of stories about how involved he was in every aspect of the company. I wanted to dig deeper and pull out more substance. In my experience, if founders were rehearsed, you just needed a bit more time to get them to speak openly. But I could tell the interview was almost over. The clock ticked down as the handlers watched closely, started shifting, and prepared to pounce. My instincts told me that we were seconds away from that point.

"Mark, what's your biggest regret as a leader over the last year?" I probed.

"That's all we have time for!" his PR person chirped, and before I could get an answer, he was gone.

The entire interview had lasted about ten minutes.

But standing in front of Mark for ten minutes was better than nothing, especially if you could read between the scripted lines. The writing on the wall said that Facebook was beginning to do some soul-searching and had updated its own status to: *It's complicated.*

While Jeff and the rest of our executives were thrilled to have the exclusive, I'd hoped we'd get more—more time often meant more introspection. But it was a start, and we had to start somewhere.

When I arrived back in New York, as I navigated the tricky combination of ambition and work, Jared and I transitioned our friendship into something more. Ethan and I were officially over, and while daily I wondered what he was doing, and often found myself doom-scrolling through his digital life on Instagram, I was sure our ending was one of the best moves I'd made for myself.

But Jared and I didn't take things slow. "Slow" wasn't a word in either of our vocabularies. Slow meant taking the time to get to know ourselves and spend time alone in our minds. Slow was intimate and honest.

We started going out until 2:00 A.M., even though I had to be at work early in the morning and he had to perform live shows. The days spun into nights woven together by a tapestry of drinks that helped us feel confident and whole, because we both felt anxious and empty. I was numbing myself out of fear—fear of being alone; fear that without an outlet to do the types of stories I loved, I was becoming a mouthpiece for Big Tech; fear that I wasn't living the life I knew I was capable of living. I was drinking too much and writing too little. The young woman who spoke to taxi drivers and enjoyed taking on stories she was proud of was fading away into the loud chatter of pings and tweets in an always-on world and a shifting media landscape.

Jared and I needed to keep going, because if we stood still long enough, the voice inside might scream loud enough to force us to listen: *You are sad, you are broken, you are afraid, you are human.* So instead, we sang to Alexa. To each other. We haunted karaoke bars, sneaking into dimly lit private rooms with cheap couches and fluorescent lights. Jared's voice was so rich, no amount of alcohol could ruin it, reminding me why he was famous. He would have a show the next day, and I would have a live shot early in the morning, but that didn't stop us. We were high-functioning and addicted to avoidance.

Looking at Jared was like looking into a mirror, except his star was brighter, his facade thicker, his walls harder. To someone who prided herself on an ability to "crack" anyone, Jared was impenetrable. I was angry that I couldn't crack him, and that he didn't want me to. I was angry that he was so easy to talk to in the chaos of that summer, yet every time we got close to anything resembling intimacy, he'd pull back. Still, I would invite him over to my apartment, where we'd fall into the sheets.

But Jared would never be my significant other, and I would never be his. Our relationship reminded me that I might never be anyone's significant other unless I slowed down and had the courage to live more honestly. But my heart felt like a balled-up cocktail napkin.

Even so, I was prepared to press *repeat* and to do it all again throughout the summer. I wanted so badly to be loved because I struggled to love myself. I was so afraid, spinning into the dark because the light is much harder to bear when you can see your shadow.

So I hid; through him, through drinks and late nights. I traveled every couple weeks, looking forward to the road. I'd listen to my favorite song, "Home," a folk anthem that was ironic, given I'd never quite found home with anyone, especially with myself. But I listened to those lyrics, *"Home is wherever I'm with you,"* as I floated through airports, counting miles and cups of coffee. Wheels up, heels down, cameras rolling.

In August 2017, Daniel and Julie were married. They rented the penthouse of the Nomad, a hotel in the Flatiron District, and I showed up, alone. Jared had gone to L.A. to work on an album, and we'd left things open-ended, noncommittal, painful.

I was single and felt like an outsider. But I had to give a speech, so I got up under the candle-lit chandeliers and white-canopied tent, and spoke to Daniel and Julie about finding a person who didn't come with a question mark. I looked at my best friend, who was beaming.

"I had this thought," I said, silently praying my speech wouldn't sound like a love-themed TED talk, "that filling your life with texture, striving for authenticity, is what makes you special and capable of the kind of relationship we always spoke about."

I continued, reflecting on our conversations at Cafe Centosette in the East Village, when Daniel and I were just getting started in our careers. Joking about finding our significant others, about our fear of settling, about finding Boardwalk.

Nearly a decade later, against the backdrop of the Manhattan skyline, I raised a glass. "To finding spectacular."

I felt completely lost, but looking at Daniel and Julie, I knew that somewhere out there, Boardwalk did exist.

In the days that followed, I wanted to text Daniel for one of our classic adventures, but I had to remind myself that he was someone else's somebody. I started using my phone like crack. Another second, another hit. Another connection. Another outing. Another shoot.

In September, I flew to the West Coast to record a couple of segments for my latest CNN special, called *Divided We Code*. The series was about how tech was beginning to tear apart politics and culture and humanity.

I met with Ev Williams in the Medium offices off Market Street in San Francisco. Everything about the space felt more muted than in other Silicon Valley offices, which had grown even more bloated than Joan the Frog in her last days.

As the crew set up, I took in the wood paneling and metal piping, passing by offices named after old typewriters. The place felt very Ev: both thoughtful and understated, with a focus on words and design. It had been months since I'd last seen him in Stockholm.

Although he'd stepped aside from his day-to-day role at Twitter, I planned on asking him about the company. I had watched as tech founders who claimed their companies were "just the pipes" tried not to accept responsibility for the content on their platforms. But that party line was getting old, as founders were beginning to engage in the delicate dance between protecting free speech and shutting down abusive content.

We sat down in iron chairs across from each other. After a bit of chat, I posed a question.

"Whether or not you like it, you have to make some decisions that are editorial, wouldn't you say?"

He took a moment to process this before responding.

"I think the fact that tech companies have to accept is that there are adjustments being made all the way down the line. There are judgments about how the algorithm works, what the system values, what the feedback loops are."

For the first time, he spoke about Facebook with disdain, calling the social network "a stockpile for junk." He attacked their fundamental business model, saying that selling ads was responsible for the growing problem of misinformation spreading across social networks.

"Facebook is filled with crap ads," he said, growing visibly agitated about what he called the "junk information epidemic." "We need to start eating some more organic, healthy stuff in order to bring reasonableness back to societies."

Generally, founders of competing tech companies didn't take shots at each other so openly. But the public had learned that a Russian troll farm had bought ads on Facebook to influence the election, and there were issues around regulation. People were being kicked off platforms for hate speech, and many were searching for alternative outlets.

The disenfranchised formed their own online communities called "red-pill forums," referring to the scene in *The Matrix* in which Neo is offered a choice between taking a blue pill or a red pill: the blue pill would allow him to continue living his life in ignorance, while swallowing the red pill would force him to wake up to the uncomfortable truth that everything he'd been taught was a lie. These forums railed against some combination of political correctness, identity politics, the deep state, and the state of the nation. Participants fanned across the right side of the political spectrum, comprising everyone from libertarians to conservatives.

As these anti-establishment forums gained traction, Erica and I interviewed conservatives in the tech world, who asked to have their identities hidden, for fear of retribution. They called themselves the new underdogs, telling me that if they revealed their political leanings, they could lose business. Reddit threads suggested they form a secret handshake. Offline, I was told there was a social contract between conservatives who met at a political event: if they were introduced elsewhere, they'd hide how they'd initially encountered one another. They were predominantly, though not exclusively, wealthy white men who believed they were being discriminated against in the aftermath of the #MeToo and Black Lives Matter movements. In Silicon Valley a culture war was playing out.

Nearly a year after President Trump was elected, the question remained: Were the same companies that promised to bring us closer together pulling us further apart? Were tech companies really the main source of polarization, or were they a mirror of the polarization that already existed? The lines between online and offline were blurring. The world I'd helped to bolster was becoming a caricature of itself. As one of my interviewees put it: "Our world is becoming a chatroom, and we are becoming our own avatars."

The political drama unfolded on cable news every day. CNN aired talking heads spouting their opinions, which were clipped and tweeted and retweeted so that people on both sides of the aisle could retreat further into their corners, filled with outrage. Panels on the Trump presidency took priority over long-form journalism, and I struggled to get airtime for my pieces on tech just as tech was seeping into society in an ugly way.

I sat in Strawberry Fields and viewed a video on PewTube, where videos censored by YouTube were given a home. The video was titled "Ovens of Auschwitz," and took aim at the Holocaust, displaying devastating images from the atrocity superimposed with cartoons depicting Jews in an anti-Semitic light against a musical backdrop of Simon and Garfunkel's "The Sound of Silence." How should I cover this? *Should* I cover it?

I decided that including the video as part of the special had value, but only if we added context. "This is where I'm running into trouble," I explained on camera, to help viewers understand the complexity of the editorial decision. "Do we even show any of this, or do we scrap it? I don't think a lot of these people deserve a platform, and clearly some of the tech companies don't either. But the news value is, they have a platform. And they are gathering, so you can't ignore it."

In the end, I also decided to interview the founder of PewTube, who secretly taped the interview and released it ahead of our special. I was growing accustomed to the ways of alternative internet circles.

I wanted to help counteract the toxic waste, to make people feel less alone, to find authenticity, but CNN had once again rearranged its teams, and Erica and I had yet to hear back about the second season of *Mostly Human*. In a professional limbo, I planned a trip to L.A. to continue shooting *Divided We Code*. Similar to our series on revenge porn and hackers, it would be turned into a thirty-minute special.

I hadn't seen or spoken to Jared in a month, but when he'd left for L.A., we'd agreed to stay in touch and "see where things went." But I knew where we stood, which was somewhere in the in-between. I texted him that I was flying to L.A., and we decided to meet up at the lobby bar of the W Hotel, where I was staying. He arrived at 10:00 P.M. on a Sunday, his only available time slot.

The W Hotel lobby bar was my worst late-night scenario—top-heavy bros formed circles of testosterone, and girls wobbled around wearing miniskirts as short as their heels were high—but I didn't want to bring Jared to my room this late in the evening. *Let me try to hold on to a shred of dignity,* I told myself as I passed a woman in fishnet stockings. But in an ironic twist, the bouncer wouldn't let Jared in because he wasn't wearing a collared shirt. The famous musician, unrecognized, didn't make the cut for the hotel's cover band on a Sunday night.

So up we went to my small room with a view of a building under construction, complete with a light switch that was stuck on dim.

At first, it was awkward. I felt like we were two strangers without anything to say, talking about nothing, laughing over YouTube videos. But after a while we were back at ease, and I remembered why we'd connected.

Eventually, he got into the bed while I stayed perched on the windowsill.

"You know, that windowsill is pretty gross. I bet tons of people have had sex on there," he said too confidently, in an obvious effort to get me to come to bed.

He lifted an eyebrow, but I hesitated, determined to break an unhealthy cycle with someone who couldn't give himself, or me, what either of us needed. I thought about how I almost hadn't messaged him that I was coming to L.A., and cursed myself for relenting. I wished I were stronger. His arm touched my arm, and I let myself go.

"At first when we met, I thought we'd date," he said, when I turned my back to him. "But we both went into the darkness together."

He wasn't wrong.

"I didn't realize your mazes were my mazes," he said. "It's like we're both locks, and neither of us has the key to the other. Our issues, our walls, are the same. We're both yin."

I get it. The part of me that knew this was our final night together wanted to say, "Enough with the metaphors, unless you're writing a song about this." But the other part didn't want it to end.

"You know . . ." he said, as we lay close. I waited for him to say something serious.

"The light from the construction makes you look pretty." He smiled.

I looked at the fluorescent light beaming through the window, setting the scene for the breakup of our non-relationship.

"I'd paint you," he added.

We lay awake for hours, talking until I closed my eyes, knowing he wouldn't stay in the morning. Sure enough, when the construction lights were replaced with sunlight, Jared said goodbye. It was at the hour before it's officially morning, early enough to let me know he wouldn't be there again tomorrow, or any other day, for that matter.

"I've got to go let out the dog," he whispered, as I kept my eyes closed.

We both knew the subtext.

A minute later, I was alone again, lying next to an imprint where his body used to be. The weight of the breakups piled up on each other. I'd spent a decade chasing big stories. I'd pushed boundaries and taken risks. I'd helped other people find their power and voice, and yet somehow, I couldn't seem to find my own. Maybe, all this time, I'd demanded vulnerability from others because I needed it most in myself.

I dragged myself out of bed to meet with Cody Wilson, whom I'd met years before when I covered his first controversial project: creating and distributing blueprints to make 3D-printed guns. His newest endeavor, a crowdfunding site dubbed Hatreon, had become a mecca for racists, anti-Semites, and white supremacists, who were getting kicked off mainstream social media platforms for terroristic content and hate speech. The site was different from red-pill forums because users could raise money for their causes, many of which were devoted to supporting racists, misogynists, and white supremacists, but I wouldn't have been surprised if there was crossover among different sites; for example, red-pill forums on Reddit, which were also known for misogynistic and alt-right ideology but didn't provide crowdfunding services.

Everything about Cody was controversial. He was currently embroiled in a legal battle with the State Department over whether his blueprints for 3D-print guns were legal. Even so, he touched on something important. He represented a group with alternative viewpoints, people who were angry at tech companies and their growing power and influence.

I left the heart of L.A. to interview Cody at a location our crew had scouted ahead of time. Against a backdrop of animal skulls, we positioned ourselves in front of his computer.

"Are you worried this kind of speech will incite violence?" I asked. I was disgusted by the content in front of me—horrific racist memes and anti-Semitic videos full of offensive imagery.

"No," he responded unapologetically. "These people are, at worst, trolls. At best, they represent elements of a political speech that should not be censored.

"I like catching interesting fish," he added. "3D-printed guns got me to hook the U.S. State Department into a lovely federal case. I'll probably be able to create a bigger turmoil from this site."

Cody Wilson was the ultimate provocateur. Trump had yet to tweet his support of Cody's endeavors around guns (that would come in 2018), and Cody had yet to become a national story because of it. But it felt like tech's roller-coaster strategy of move-fast-and-break-things innovation under the guise of "trust us, we're making the world better" was now approaching a dangerous juncture. As I left the skull-bedecked apartment, Cody's words reverberated in my mind.

"The closest place my politics come to any traditional school is anarchist. I don't like the imposition of state controls over human flourishing, creativity, freedom, individuality," he said. "So the way to oppose these things is to undermine the powers of traditional liberal institutions."

Cody may have been on the fringes, but he represented a contagious way of thinking. He hid behind the First Amendment in order to say and do anything. He created chaos, spread hate, and incited violence, and then pointed to the Bill of Rights. By fashioning himself as a champion of freedom, he positioned anyone who opposed him or his actions as an enemy of freedom. This included any person, company, or institution that sought to combat hate speech, hate crimes, and disinformation.

Ironically, even in the fine print of their community standards, companies like Facebook and Twitter and Instagram had used that very same argument. How could they censor posts when the platforms were based on the premise of free speech? But while they claimed to take a noneditorial role, they were secretly playing traffic cop, directing ads and information to have the maximum impact. By actively manipulating who saw what information, tech was manipulating the world—and both users and foreign governments were exploiting that fact.

Big Tech was caught with its pants down. It was only a matter of time until politicians started paying attention. But to what extent were they responsible for fixing the problem they'd, in part, created? Was that even possible? Tech companies responded with an attempt at self-policing—and it wasn't just the anarchists who were taking issue with it.

CHAPTER 17

All Hell Breaks Loose

Erica and I sat in our secret room writing our 2018 New Year's resolutions. Our lists throughout the years had a lot of check marks. *Build a tech beat. Create a special. Create a show. Interview Mark Zuckerberg.* I knew what I was supposed to want next: to become an anchor. But that goal didn't fit my ethos. I was too deeply entrenched in reporting from the field and in nuanced long-form storytelling.

Unlike in prior years, when I'd volunteered to work through the holiday weeks at the end of December, I went home to visit my parents for a week and then hopped on a plane to Mexico City with a friend to spend New Year's Eve surrounded by music, culture, and food. By the time I returned, I was recharged and ready to take on the year. But as soon as February arrived, Erica and I were slammed with company news, and this time it was personal.

The digital overlords had decided to dismantle the tech team. During a company-wide meeting, the changes were couched as another "restructuring" that was needed to "consolidate digital." In other words, digital was under pressure to make back the money it had spent, and the cuts had to come from somewhere. Like any bad news at a big company, the reorganization was presented as an opportunity, but what it really meant was that the tech team we'd spent years assembling would be broken apart, and everyone Erica had worked hard to hire, and those I'd worked with, would be scattered to different teams. I was shocked.

The changes would come as CNNMoney eventually transitioned into a new unit called CNN Business, which would be announced later that year. Ironically, after the tech team was dissolved, one of the main focuses of the unit would be Silicon Valley.

Worst of all, Erica would no longer work with me. She would be in charge of live content online, which essentially meant producing Facebook Live segments. I would work with producers who were my junior.

"Did that just happen?" I asked Erica as we stepped onto the elevator.

"It did," she said, her voice quiet and measured. Immediately, I knew that no matter what type of corporate glaze they put on it, this was the beginning of the end for both of us.

Dumbstruck, I walked out of the office and marched toward Central Park to surround myself with nature and nineties grunge music. I could smell chestnuts roasting and hear street vendors calling out to tourists as Everclear's "Santa Monica" blasted in my ears. I remembered my first day on the job, looking out the window, taking in this view of Columbus Circle and the park. The view, these smells and sounds, had become part of my DNA. But I had grown since then. And no matter how wonderful the view, if I couldn't keep growing, a braver version of myself knew I'd need to leave it all behind.

After a decade of creating a unit, it was hard for me to transition to working with producers who had a different vision. Pieces needed to be *shorter.* My editorial guidance about ensuring we include nuance in interviews was dismissed by producers junior to me, who had never covered technology and who had different priorities. Increasingly, I couldn't hide my frustration. To make matters worse, Erica and I officially received word of what we already knew: *Mostly Human* wouldn't be given a season two.

While we struggled to adjust to the changes, a breaking-news bombshell was about to drop. On March 17, 2018, the world changed. Just a day before, almost no one had heard of Cambridge Analytica. But when a pink-haired data consultant blew the whistle on the company's tactics, the news started trending across the globe. Donald Trump's campaign had hired the consulting firm to use the harvested personal data of more than 50 million Facebook users, without user consent, in an attempt to influence the 2016 presidential election. Americans expressed their shock and horror on morning talk shows and on social media. But nothing could be done. The researcher who gave Cambridge Analytica the data had obtained it legally, in compliance with Facebook's user terms (although Facebook said that the researcher violated its rules by giving the data to a third party).

How had this happened? How had Americans been duped? How had we, willingly, created Facebook accounts, complete with long consent forms written in small print, that allowed people building apps to access more of our information with less of our knowledge?

Emergency push alerts splashed across cell phones, forcing us to see that we'd given away our privacy for the price of likes, and that we were being manipulated in hundreds of unknown ways. An innocent-looking personality quiz was a potential gold mine of information. Researchers could dig into not only that, but our friends' data as well. As we frittered away hours posting pictures of sunsets and margaritas, our innermost secrets were up for grabs. The more we clicked, the more data researchers could license to firms like Cambridge Analytica, who developed software to analyze users with the scrutiny of a Freudian psychiatrist. Behavior turned into patterns, and once we became predictable, we could be influenced. Or at least that's how Alexander Nix, the CEO of Cambridge Analytica, billed the methodology.

The company called it "behavioral microtargeting." And while the idea of using psychological traits to influence the behavior of voters wasn't groundbreaking, Nix positioned himself as a guru in the method of digital ad targeting.

Though many people still had questions about how effective his software was, and how much anyone could actually *do* with our data, Nix helped the Trump campaign target people very specifically and personally. Once the news was made public, the idea that our data could have been employed to manipulate us scared the hell out of everyone.

This "WTF!?!" moment moved the conversation about data collection and privacy from tech circles and elite media into the mainstream. Journalists across the world charged at the story. The *Guardian* published an investigation revealing that Facebook knew researchers had been obtaining and using data without permission since 2016, and that Facebook had never followed up on its request for Cambridge Analytica to destroy the data.

Suddenly, it seemed as if Americans had been hijacked. The story was personal. Facebook had asked users to be transparent about their whole lives, to share more, to connect more—but users hadn't been awarded the same transparency. A feeling of betrayal ricocheted around the globe. People demanded to hear from Mark Zuckerberg. They needed to know: What the hell was Facebook doing with their data?

It was the topic of the morning at CNN. Jeff Zucker sat in a swivel chair at the head of a large oval table in Strawberry Fields, waiting for his news executives. By nine, they were gathered around him, in order of importance, ticking through the headlines of the day. Facebook was front and center. The Cambridge Analytica scandal was gaining steam, but Mark Zuckerberg and Sheryl Sandberg were nowhere to be seen. The public's outrage was snowballing, and every reporter was trying desperately to stay ahead of the questions, but no one had any answers. Amid the frenzy, Jeff asked a question that would be delivered to me in minutes: "Is Laurie Segall on this?"

I received word from one of our planning producers: *You have a go on Cambridge Analytica*. I knew the story had legs—major ones—and I sensed that this was a historical turning point, both for tech and for me. In a sense, I'd been working toward this moment since my earliest days in the city a decade ago.

I remembered my segment eight years prior with Facebook's Head of Product, Chris Cox, as Facebook turned ten, and his fear that the company wouldn't maintain its ability to stay scrappy. Looking back, I wondered if that fear was grounded in the sentiment that as Facebook became more of a behemoth, it would have to fight harder to hold on to its initial vision of connecting the world. I recalled the next time I'd visited the Facebook campus after my interview with Chris. New buildings had popped up, more industrial, more impersonal. The engineers in charge of growing Facebook, I was told, got better real estate than the security team in charge of keeping the platform safe. Expansion was the name of the game, and no one could imagine pumping the brakes. But now, it had all come to a head. And the company's most prominent leaders were MIA.

There was a race to get to Zuckerberg and Sandberg. Like every other journalist, I was digging deep into my Rolodex. Placing call after call, I paced through the newsroom, outside Jeff's office, by the set where a live show was taping. I hunched over my laptop in any open seat I could find. When a story was coming in fast, I liked to move with it: walking around on calls, sitting at empty desks and in corners of the newsrooms to take notes. And chatter was pouring in. Multiple sources whispered that there were internal discussions at Facebook, and growing resentment within the company's ranks. Employees grumbled that Mark and Sheryl were being protected by their own filters, that their people and their people's people were worrying more about their images than those they served. Even Facebook management was simmering with resentment. Where were their leaders at such a pivotal moment? The scandal wasn't blowing over; it was blowing up.

I had enough information to piece together a picture of what was happening inside Facebook, so before shutting down for the evening, I filed a story. The following morning, there was an email in my inbox: *You're on-air with your story in twenty.* I raced to hair and makeup, heart thumping, feeling the rush of excitement I always got when I had exclusive information that would push the needle. Between a flurry of brushes and Wonder Wedges, I got another email: *James will be joining you for the segment.* I took in a sharp breath, coughing on a cloud of powder. "James" was a classic, square-jawed TV personality who loved to hear himself talk, even when he wasn't saying much, which was most of the time. *Ugh. Why?* Gritting my teeth, I tapped out, *Great!*

Minutes later, I was on set greeting the anchor, who was a self-proclaimed "woman who supports women." I sat next to James: me with my actual news, him with his airy pontifications.

"James, let's start with you," she began, knowing full well that I was the technology expert. As James spoke loudly, gesticulating, conducting an invisible orchestra with his ego, I sat silently, patiently, politely. *I'll be concise and specific,* I told myself as he droned on. *I actually know what's happening inside Facebook. I don't just have opinions; I have sources.* Agreeably blank faced, I waited for him to finish sharing his newly formed opinions on social media as the brief segment's valuable minutes ticked away. With about thirty seconds left, the anchor finally turned to me.

"Laurie, really quick: What are people inside the company saying?"

I got in about two sentences before she smiled and said, "Well, we've got to wrap."

When we finished, everyone shook hands while I sat like dead weight, wondering how this blowhard had managed to own the moment. This was *my* beat. How was it that there was always an entitled guy with less experience who was taken more seriously, who spoke over me and got a seat at the table?

Unclipping my lapel mic, I thanked everyone on set, fearing that if I didn't smile, I'd scream. James pushed past me, looking down at his phone, tapping out an email with an aura of self-importance. Watching him go, I made a decision. Instead of bitching to Daniel or Erica—instead of waiting to be trampled on by the next suit—I would make my move. As I walked off the set, I had one thought: *I'm booking Zuckerberg.*

I didn't tell anyone what I was about to do. Fueled by outrage, I had a calm sense of clarity. Mark Zuckerberg would eventually speak, and when he spoke, it would be with me.

I pulled out my laptop and got ready to send a message to the founder of Facebook—on Facebook. I searched his name and saw that we shared more than a hundred friends in common. Many were his colleagues, some of whom had moved to other companies, and others who'd remained in his orbit. Given how many connections we shared, I knew that the Facebook algorithm would look at the data and not categorize my message as "random," sending me into the purgatory known as the "other inbox."

I drafted my message, explaining that I understood there was a lot happening behind the scenes. "I can't tell you how important I think it would be to hear from you as a leader," I wrote, asking him to consider doing something on camera with me. When I finished, I took a brief look at my long-winded paragraph and thought, *Segall, do not overthink this.*

Then I pressed *send.*

For the next few minutes, I reread the message over and over, knowing that whatever happened was out of my control. Trying to forget what I'd just done, I clicked through my emails, sorting through the various network requests. CNN, HLN, and CNN International asked me to appear on-air to answer the question: Where is Zuckerberg? I asked the same question as, every twenty seconds, my eyes flicked back to Facebook.

Finally, after forty-two minutes, I saw it. Mark Zuckerberg's icon popped up. He'd opened my message.

Okay, I thought. *Maybe we're getting somewhere.*

Afraid to leave my seat for fear that I'd miss out on one of the most important interviews of the year, I decided to attack on all fronts. Mark had seen the message. But I realized the person who'd help guide the decision was his rep. I called Facebook's head of PR, Caryn, a no-nonsense kind of person you'd want to bring to war with you. We talked for a bit, and I told her that I was the person to do the interview. Phone to my ear, I looked up and caught sight of the ever-present CNN news crawl. A winter storm was coming; before long, all flights out of New York City would be canceled.

"I have to know if I need to be on a plane," I said to Caryn, trying to convey a sense of urgency.

She said she'd call me back.

After checking one more time for Mark's response—nothing—I jumped up from my chair and ran-walked to a seventh-floor studio, where I prepped for an interview with a larger-than-life British anchor named Richard Quest. As soon as we were live, Richard banged his hands on a glass table and boomed, "WHERE ON EARTH IS ZUCKERBERG DURING ALL THIS?"

When we cut to the break, a call came through. It was a Facebook rep.

"Tell me something good," I said, already at the elevators.

"It's looking good," the rep replied. "No promises."

Back in the fifth-floor newsroom, I briefed Jack on the situation, telling him that if this came through, we were going to need to find a flight out of New York before the storm came barreling in. As he took notes, I got another call.

"Get on a plane," Caryn said.

I sent out two emails in quick succession. The first went to a small group of need-to-know people who would disseminate the information: *I got Zuck.* The second went straight to Jeff Zucker: *Erica will be producing on my end.* I looked out the window; the clouds were already a dense blanket of Gotham City gray. We had to move quickly.

As I headed home to pack my bags, Jack conferenced me in with CNN's travel department. Phone jammed between my cheek and shoulder, I hurried into my apartment, snatching tubes of makeup and throwing them into the smallest bag I'd ever packed. I was blindly pulling a dress out of my closet when the nice woman from CNN travel said that there was a storm coming, and would we consider flying out the next day? I declined and offered up my firstborn if she could make this work for us. The travel lady put us on hold.

You're scaring her, Jack texted me.

I grabbed a toothbrush. *Anderson Cooper would be on a private plane by now,* I thought bitterly, though aware that this wasn't entirely accurate.

When the lady returned, she'd found a miracle. There was a flight out of Newark in the next couple of hours, one of the last to make it out. We would connect in Oregon in the middle of the night, and from there, fly to San Francisco, where we'd have just enough time to change clothes before racing to Facebook's campus. My bag half zipped, I called an Uber.

I arrived at the airport with thirty minutes until final boarding. The long security line was winding through the corridor, and somehow, in my hour-long call with CNN's travel department, my TSA precheck number hadn't made it through to my ticket. I did the mental calculus: if I waited on the general boarding line, I'd miss my flight.

Frantically, I pulled out my cell phone. I searched through my emails, apps, and key chains, but I couldn't find my precheck number. My breathing was heavy, my clothes sticking to my body. I was about to break through the barricade when I realized there was one person in the world who could save me. I texted Ethan.

Hey, how's it going? I typed.

I watched as three dots came and went. Ethan was deciding how to respond, figuring out my angle. We hadn't spoken in months. Did we need to do pleasantries, or could I just be transactional? God help me, I cut to it.

I'm about to interview Mark Zuckerberg. Please keep between us. Any chance you have my TSA precheck number?

I closed my eyes. *I was awful.*

When I opened them, a second later, he had delivered. He was so organized—he deserved a medal, really. He just wasn't the One.

I got through security and started sprinting. I saw the back of Jack's head, weaving through the crowd, his tripod wobbling next to him. Five minutes later, we were at the gate, joining the final boarding. Both of us looked like we'd been in a minor dispute with a tornado. A strand of sweat-slicked hair snaked across my cheek as I looked down at my phone. The emails were beginning to stack up, the wheels on the CNN news-making wagon set into motion.

I shot off a message to Erica, who'd always bolstered me during moments of chaos. *Made it.*

She texted back immediately. It was her son's first birthday, but instead of watching him stick his hands into an oversize sheet cake, she was on the phone with CNN travel, conducting her own debate to reroute another producer midair, midstorm, from Philly to Menlo Park, while simultaneously trying to convince Caryn from Facebook PR that having a CNN Live truck near their campus wouldn't be the end of the world. Erica could have had an alternative career as a hostage negotiator.

My cell service cut out in the jet bridge dead zone, where I was finally able to reflect on the gravity of the moment, and whether I'd packed any underwear. I shuffled onto the plane, lugging my carry-on past business class, past premium economy, and past general economy until I got to the very back, where my middle seat awaited me. As I struggled to cram my bag between two oversize rollaboards, the guy beneath me screamed abruptly and clapped his hands. I looked around, worried something was wrong. But no—he was watching a basketball game as if he were at a bar with his best friends. Dodging a rogue elbow, I popped in my headphones, and soon enough, the whir of the engines drowned out his grunting and cheering.

I worked for two hours straight, reviewing questions, flow, setup. The distractions around me disappeared—until my seatmate unwrapped a piece of salmon, confirming that I must have murdered someone in my past life to deserve this seating arrangement. I tried to refocus, but the fishy smell lingered. For the remaining four hours, my innards felt like a pair of socks in an otherwise empty washing machine.

When we landed in Oregon, I collapsed at an empty coffee shop next to my gate. I pulled out my questions to review, and barely had time to look over what I'd written and order caffeine before I heard over the loudspeaker that my flight to SFO was boarding.

A quick one and a half hours later, Jack and I were in San Francisco, slumped over in the back seat of an Uber. Just minutes away from our long-awaited freshen-up, we heard a siren behind us. Illuminated by flashing red and blue lights, I wondered who the police car was chasing, until our driver slowed to a stop, and the cop car did, too. My sleep-deprived brain went into hyper-speed. *Surely, I can charm us out of this*, I thought, sitting up straight, preparing to negotiate. I put on my brightest smile—and then caught a glimpse of myself in the rearview mirror. I had deep circles under my eyes, and my hair was beginning to spike in a way that had inspired an ex-boyfriend to nickname me "Sonic the Hedgehog." I realized that between the lack of sleep, the airport chaos, and Salmongate, I had about as much charm as the Loch Ness Monster.

I hunched back in my seat and let justice take its course. The cop wrote out a speeding ticket and I spent the remaining six minutes of the ride apologizing profusely to the driver on behalf of the entire Silicon Valley community, where the rich were becoming even richer, while he became poorer.

Finally, we arrived at the hotel, and there, waiting, I found Meagan—my makeup fairy godmother. I'd first met Meagan years back, during one of my many travels to San Francisco. She was based in the Bay Area, and since then, she'd become my go-to makeup artist. She'd given me fake eyelashes for some of the biggest interviews of my career, and had witnessed many newsroom dramas, founder freak-outs, and self-inflicted panic attacks. As a perk, she'd taken up Reiki and exuded an incredible sense of calm. I wrapped myself in her aura and followed her into my room.

As Meagan unloaded her various tools, I unzipped my carry-on. Between the call from Caryn and the race to the airport, I'd haphazardly grabbed a fistful of clothes, which meant that my options were . . . limited. I stared down at a black skirt with an angular slit and a sleeveless black knit shirt with a red and orange turtleneck. I took a breath. The interview would appear live on *Anderson Cooper*, and CNN was promoting it like crazy. It would be the biggest exclusive of my entire career. This outfit—well, it was "edgy." Meagan called my name, and I picked up the sleeveless turtleneck. *Screw it*, I thought.

By the time we arrived on the Facebook campus, I was approximating a professional. We were ushered into one of the newer buildings, through a lobby with colorful graffiti splashed over soaring cement walls. We registered, received our badges, and passed by the Facebook "sign" walls that displayed positive messages, like "Empathy." The signs that read "Move fast and break things" were long gone.

Despite the chaos on the outside—the swirl of media, users, and politicians demanding retribution—it was strangely calm on the inside. Facebook's handlers stuck to us like shadows as we walked through rows of open desks and whiteboards. Meagan followed me with her makeup bag, Jack close behind, rolling our small setup. We stopped in a nondescript conference room and in a corner, I saw a small device that looked remarkably like a tape recorder. The subtext was loud and clear: nothing we did was private.

Our crew spent an hour and a half setting up. We were told that Mark was particular about the chair he'd sit in, which made sense. He wanted to feel comfortable in what was arguably one of the most uncomfortable moments in the company's history. I was just thrilled we'd be sitting.

When it was finally time to start the interview, the temperature in the room had dropped by ten degrees. I should have remembered my experience in Chicago, which I'd dubbed the "refrigerator interview." Now I was regretting my sleeveless knit top.

As I tried to relax, Mark entered the room. The energy was different from when we'd first met. Our last segment had been a carefully curated PR op devoted to emphasizing the power of Facebook groups and the community the company was building. We'd stood around with beaming members of groups who had found one another on Facebook and were meeting at the summit. But that energy was long gone. Now there were few people around, and lots of explaining to do.

"Hey!" he said, approaching our crew and starting a conversation. He seemed less guarded than he had the last time we met, speaking more freely. But I could sense his nerves; in this moment, he seemed strikingly vulnerable. After Meagan applied a bit of powder, we took our seats to start the interview. But within seconds, Mark stopped, asked for a minute, and abruptly left, followed by his handlers.

I sat there motionless, other than shuffling my notes and signaling to Jack to gather intel, while we waited for an update. Our so-far-nonexistent segment was currently being promoted to air later on Anderson Cooper's show, with CNN's countdown clock ticking off the minutes. Was the interview still going to happen? After ten minutes, the handlers returned with an update.

"Do you mind if we switch rooms?" one asked. "It's not cold enough in here. Mark would prefer the room a bit colder." It could have been his history of sweating profusely during major interviews, or the all-around intensity of the day, but it was clear Mark felt more comfortable in sub-zero temperatures. I did the mental calculus. We needed to get the interview started to make airtime. If moving to Siberia was the key to getting immediate answers to the questions I had scrawled, we'd move.

We scrambled to set up in the new, more-frigid room, and this time prep took only twenty minutes. There were no chairs, only a small couch. I took a seat on one side, and eventually Mark returned and sat next to me, knee to knee.

"You sure you want to be this close?" I joked. But the smiles were gone. All of us were ready to do this.

We started the countdown to ensure our mics were working correctly.

Zuckerberg. Count to five.

"One, two, three, four, five."

Segall.

"Five, four, three, two, one."

And we started to film.

"Mark, what happened? What went wrong?" I asked.

He stared at me for about a second too long, and then ticked off his scripted response.

"I'm really sorry this happened," he said, making sure the apology came first. He went on to explain the details of what went wrong, and how Cambridge Analytica exploited outdated practices.

It took a minute, but we found a rhythm.

"Should Facebook be regulated?" I asked.

"I'm not sure we *shouldn't* be regulated," he replied.

I asked him whether he would testify before Congress. He evaded the question, saying that there were people better qualified to answer specific types of questions. I challenged him, emphasizing that people wanted *him*, the creator of Facebook, to show up. He wouldn't commit, but he left the door open.

As we approached our allotted twenty-minute mark, the Facebook handlers tried to cut us off, but Mark kept going. We continued to talk for another ten minutes, then another. When it seemed that there was nothing left for him to say, I asked him one final question: Would he want to build a kinder Facebook for his children?

Mark let out a breath. Something almost imperceptible shifted in his eyes. Maybe it was because the interview was drawing to a close, or maybe it was the significance of this particular moment, and what it meant for the future of his company. Or the mention of his children. I looked closer. He had started tearing up.

"Having kids changes a lot," he said.

"Like what?" I asked, avoiding eye contact with his PR handlers, who were long past shifting from side to side and were now ready to pounce.

"I used to think that the most important thing by far was having the greatest positive impact across the world. Now I really just care about building something my girls are going to grow up and be proud of."

"Do you feel like you're doing that?" I pushed. The question hovered, suspended over the carnage of the election manipulation, the data collection, the growing concerns about tech's effect on our mental health—on the overall health of society.

"I do." He spoke slowly, measuring each word. "We are committed to getting this right for people."

There I was, sitting across from an engineer who'd built the largest social network imaginable, who was able to do that by seeing the world in nuts and bolts, in ones and zeros. Empathy and a deep understanding of human beings weren't necessarily part of that equation. He had created his company when he was just a kid, long before he'd had children of his own, long before his daughters disrupted his own internal algorithms. The idea was to connect the world—for money, or power, or social good—but his blind spot seemed to be his own inability to connect. He was guided by optimism and a fair amount of hubris, along with a filter that had surrounded him for most of his adult life. The company had become a behemoth, and now its business model, its role in society, and its executives' intentions were being questioned.

It's an odd feeling when you sit across from someone at a moment that you know will resonate in the future. Everything in that office felt weirdly small on a day so big—the room where we had the conversation, the limited number of people present, our cameras. There was no elaborate setup, yet I knew the interview would be picked up everywhere.

Perhaps most interesting to me was how human Mark was. We had officially crossed over into a new era where our tech giants, the people who had been made into gods in the boom years, were being forced to reenter the stratosphere and make their way back down to earth. The interview was a dance, but the moment was one I knew would cause aftershocks down the line, transforming the narrative around not just Facebook's impact on society, but also technology's overall impact on us mere mortals. The Wild West days of blind optimism and promises of connectivity without consequences had screeched to a halt. We were now wading knee-deep into a new era of scrutiny and messy complications, and the stakes were high.

As we sat knee to knee, I thanked Mark for his time, and then he was gone. The twenty-minute interview had lasted forty minutes.

Before the door closed, Jack and I were back on our feet, packing up and racing out of Facebook's offices to feed our tape to a live truck located somewhere in Menlo Park. The footage was beamed to the CNN newsroom at lightning speed and minutes later, the interview aired on *Anderson Cooper 360°*.

The segment was picked up around the world. Almost immediately, newspapers ran op-eds debating the power and influence of Facebook. Talking heads weighed in on technology's role in protecting user privacy, and Twitter blew up with people parsing Zuckerberg's words. It was the first time he'd ever said that Facebook maybe, just maybe, should be regulated.

When I went live on *Anderson* hours later to offer my own insights, I was joined by a panel of men. They dissected my interview, leaving me little time to share my opinions. But I smiled. At the end of the day, no matter how many guys were shouting over me, I was the one who'd gotten Mark Zuckerberg.

CHAPTER 18

McScam

I knew it before she said the words. I could read the color of her cheeks, an extra shade of pale with a dust of rose, and her arms crossed tightly over her chest. She was holding back something important, struggling to find the words. So I said them for her.

"You're leaving."

Erica nodded. She was silent. I didn't know you could feel so much at once: immense nostalgia before someone you love walks out the door; an overwhelming pride in her bravery to walk; and an undertone of envy that she had the bravery I lacked.

I could sense that she was nervous, so even though every part of me felt like screaming, "Don't go!" I said softly, "I get it."

"NBC," she said. "They're launching a streaming news channel. I'm going to help run it."

With Erica's experience creating CNN's first streaming show, *Mostly Human*, she was the perfect fit.

"I'm so proud of you," I whispered, reaching out to hug her as I pushed aside my fear of remaining at CNN without my other half.

Nothing would be the same. But if I was being honest, everything had already changed: the nature of cable news; the political climate; our new roles in separate departments. We'd spent years playing the corporate game of chess, but there was only so long you could play on the same board.

"It's time." My words were directed to her, but I meant them for both of us.

"I know."

We sat in silence in our secret room, gazing out at Columbus Circle, reflecting on the early mornings and late evenings. I thought about my first stand-ups, looking into a camera and doing everything to try to make those live shots feel natural. She'd stood by me like a parent watching a toddler learn how to walk. She'd been my backbone when I needed to stand straighter and believe in myself. I'd helped her heart beat faster, pulling her into the tech beat, into years of adventures. We'd both lifted each other.

It felt like we'd gone to war together, too. Our battleground was the newsroom. We were soldiers for a good story. We'd covered kidnappings, internet terrorists, and bombings, running toward live shots. She'd whispered through the IFB into my ear so many times that when I doubted myself, the voice pushing me forward was hers. We were ruled by the belief that if you had heart and kept fighting, if you figured out the right doors to knock on when others inevitably shut, you'd make it.

We had never held each other back, and I wouldn't start now. Not only did I have to let her go, but I planned to be her rock, no matter how ambitious our checklists grew. I would always remind her how special she was.

The email came a month later.

Segall,

It's pretty much impossible to put into words what our relationship has meant to me. Ours is a true example of the whole being greater than the sum of its parts. And . . . if you think about the totality of insane things we've done together—chatting with ISIS, reporting from Ariel Castro's basement, a swinger's party, and some very sketchy basements with sex workers; getting our car hacked at Black Hat; getting a revenge porn hacker to admit his crimes . . . on camera—it's quite a miracle that we're both alive and disease-free.

Gushing isn't something that can be checked off in the planner, so I've taken the liberty of starting the new list instead. I expect all items to be checked off in short order.

LAURIE: Continue telling the most powerful stories that people don't know they'll end up caring so deeply about, with the empathy that only you bring
ERICA: (Re)invent streaming news
LAURIE: Start an empire
ERICA: Miss working with Segall terribly

. . . To be continued over drinks, monthly power dinners, and anytime you need a smile.

I'll never be more than a text away.

She signed off simply: *E*. It was the last email she sent as a CNN employee before her email address was shut off.

I'd been through breakups, but Erica's departure opened a new corner of introspection I didn't know existed. In the vacuum of her absence, I allowed myself to slow down enough to think, to let my own three dots linger. I wished Erica all the success in the world, but what did success look like for me? What kind of "empire" did I want to build?

I kept returning to a larger ambition: to create a media company devoted to finding authenticity in a filtered world. I wanted to establish a space where I could build ideas and projects like *Mostly Human,* where I wouldn't be told I was too aggressive, or ambitious, or ungrateful. I wanted to bolster other voices that I thought were important. I didn't want to wait patiently for someone in a corner office to understand my vision. I wanted to build my own company. But how to do it? I had no idea. I'd spent years covering startups, but when it came down to it, I wasn't sure I had it in me to start my own.

I thought back to what Chris Sacca had said to me in 2016: "This is a very special, different journey that is not actually suited to everybody. You have to be a little fucked up, you have to be a little weird, a little crazy."

I started writing notes on napkins, bullet points on Post-its. I continued to keep my head down at work, but I allowed myself to dream.

"Is there anyone you haven't booked?" the anchor joked on-air. "Who's next? The pope?"

I gave a little laugh, pushing aside an imminent feeling of dread. I had been on the go for a new series called *Human Code,* which was a sequence of interviews with major tech CEOs. I'd just finished with Apple CEO Tim Cook, who'd announced a new tool to curb tech addiction. Landing Cook was a huge get, but the interviews were nonstop, short-form hits that made me yearn to dig deeper into the issues.

CNNMoney was officially being rebranded as CNN Business, and I'd been tasked with coming up with a splashy series for its launch.

Over the coming months, I flew from coast to coast, sometimes once or twice a week, meeting founders for interviews skimming the surface of ethics and the future. In San Francisco, I met with Ben

Silberman, Pinterest's CEO, a soft-spoken former engineer who revealed that he was worried the company was creating a culture that put too much pressure on fantasy and prevented people from creating experiences in real life.

I talked to Anduril founder Palmer Luckey at a compound outside L.A., where he and his team were building weapons of the future and defense technology, taking on important questions such as: Should AI be allowed to decide whether or not to kill? And should drone technology be used to track down people illegally crossing the U.S. border?

When I spoke with Salesforce CEO Marc Benioff, he compared Facebook to cigarettes in terms of its addictive quality and called out the company's practices: "When everyone and everything is connected, you're going to have to really think about, do you trust what is happening?"

I spoke to Uber's new-ish CEO, Dara Khosrowshahi, who'd taken the helm after Travis Kalanick was pushed out. The former Expedia CEO was specifically brought in to deal with hard questions.

"What would you say to Susan Fowler?" I wanted him to address the woman whose bravery and inability to accept the status quo were beginning to create change. After all, it was her blog post outlining the alleged sexism at Uber that sent shock waves through Silicon Valley.

"I would say she did good," he said, seeming both thoughtful and uncomfortable. "Listen, there's a pendulum swing right now, and I think what's happening is incredibly important; it's ultimately going to be good. But when you're going through periods of change, there's going to be pain."

I felt that in every sense. The ethos of the Silicon Valley I'd fallen in love with was shifting. The outsiders were now the ultimate insiders; I was, too. But the more I sat in front of the tech titans who would take us into the next world order, the more I felt an overwhelming desire for more—more long-form storytelling, more transparency and authenticity, more nuance.

During this time, Silicon Valley had started familiarizing itself with Washington, D.C. For as long as I could remember, tech founders had viewed Washington as slow, corporate, and full of red tape. There had been a clear separation between tech and government, but as anger built across the country, that changed fast. Now tech companies were forced into the hot seat, facing existential questions about their platforms. Tech leaders were grilled by politicians with partisan interests and had to defend their companies against charges that they were biased against conservatives.

The first hearing to garner buzz came a month after I interviewed Zuckerberg during the Cambridge Analytica scandal. He'd conceded he would appear before Congress "if it's the right thing to do."

In April 2018, I flew to D.C. to watch him make good on the statement and testify.

Outside, people dressed as Russian bots held protest signs, waiting to catch a glimpse of Zuckerberg. Inside, extra chairs were brought into the chambers to make room for all the interested parties attending the hearings. The tech writers who sat with me at tech conferences that mainstream media rarely paid attention to saved seats for each other on the Senate floor. The room was buzzing, awaiting the CEO of one of the world's most hyped—and in the last months, hated—companies, who was there to answer to politicians.

In August, Jack Dorsey called Sean Hannity to defend the company against accusations of "shadow banning" conservative content, which involves limiting the visibility of certain content on the platform so users have a harder time finding it. At a separate hearing on Capitol Hill in September, Dorsey had an exchange with Senator

Mark Warner about whether people had a right to know if they were being contacted by a human or a machine. I thought of Laurie Bot, about the ethical considerations that loomed over the future as the lines between humans and machines blurred.

We were officially entering a new era.

The long walk from JFK Airport Gate 25 in Terminal 4 was becoming as routine as brushing my teeth. I was exhausted. After over a month of constant travel, my back was aching, and my soul was seemingly lost at baggage claim. I could hardly grapple with the decision that had stalked me across the country: my contract at CNN was about to expire in three weeks, and I had to decide whether I was staying or going.

When I arrived back at the office, I made my way over to Edit Bay 3. Ross was always good for decompression conversations; we had a decade of them under our belts. He was squinting at the screen when I walked in.

"This is so crazy," Ross said, shaking his head. "McDonald's is being sued for that Monopoly game thing they did a hundred years ago."

"What?" I leaned over his shoulder to read the headline: "How an Ex-Cop Rigged McDonald's Monopoly Game and Stole Millions."

My throat closed. I could feel the color drain from my face as I sat carefully in the swivel chair next to Ross, scanning the article.

Apparently, there was a McSting around the supersize Monopoly game, and the entire thing had turned out to be a McScam. The most prized properties had been distributed to a network of insiders, including mob bosses, strip club owners, and drug dealers.

According to the article, a former police officer who was director of security at the marketing agency responsible for making the

Monopoly pieces had rigged the system so that the highest-valued pieces—i.e., Boardwalk—went to people he knew.

Staring at Ross's screens, I let the weight of it sink in. I'd eaten hundreds of meals at McDonald's. But no matter how many times I'd peeled back the sticker on the supersize fries, hoping for an escape from my childhood, I had been doing it in vain. The whole concept was a total fraud.

The evenings I'd spent toasting Boardwalk with Daniel were all based on this rigged Monopoly game that had contributed to fifteen pounds of unwanted weight and a theory I'd adhered to my whole life.

"Ross, I don't think you understand what a big deal this is," I said slowly.

"Are you okay?" He looked at me with concern.

Boardwalk had never existed. It was nothing but false hope. No one had won a million dollars. It was simply unattainable, in a highly caloric, unjust way.

I couldn't breathe. I'd spent too many years chasing Boardwalk; drunk too many martinis with Daniel discussing how to attain spectacular. My entire life philosophy was built on the idea that if you fought hard enough, if you refused to settle for mediocre, if you held out, it would come. Even in the tech realm, I lived and breathed the idea of Boardwalk—that you just couldn't settle. It was how empires were built, industries were disrupted, and the world was transformed.

I was holding out for Boardwalk in every aspect of my life. But what if it was all a scam? What if Boardwalk didn't exist?

I immediately texted the article to Daniel.

Oh god, he responded.

I stood frozen with the phone in my hand, Ross's voice somewhere beside me. I didn't care that I never won a million dollars playing McDonald's Monopoly. But the idea that I'd never had a shot at achieving what I thought was possible turned my world upside down.

What if you refused to accept the status quo; you went through all the mental torture, the breakups, the hard career moves, the internal and external battles, to fight for the idea of something better—and it just didn't exist? It was all a McScam?

CHAPTER 19

Boardwalk Is Within

The decision to stay or go lingered in front of me like a terrible first date. I just wanted it to be over. I was proud of myself for trying, but it needed to be over so I could get to where I eventually wanted to be.

My first thought was to continue working with CNN, but to make the relationship nonexclusive. Staying in some capacity would provide security while I went off to create a company of my own. It was a rosy idea; however, as in any nonexclusive relationship, there was a "but." The CNN bigwigs didn't want to let go. They wanted an arrangement with strings—first looks, exclusives, and so on—which felt too constrictive. Which meant I needed to stay or go. I had until 5:00 P.M. the last Monday of November 2018 to make a decision.

As the deadline approached, I was racked with anxiety. I'd wake up in the morning before the sun came up, often listening to the familiar cranking of garbage trucks making their rounds, staring at the ceiling with the same thought: *What if I had the nerve to go?*

I could easily tell my own story from a thousand-foot perspective: *Laurie, this is the climax! It's where you make the decision that becomes a turning point in your life.* But like Steve Jobs used to say, "You can't connect the dots looking forward; you can only connect them looking backwards. So you have to trust that the dots will somehow connect in your future."

The problem was, I was too entrenched in fear and anxiety to trust in anything, let alone myself. Clarity was earned, and I was too busy running to stand still long enough to find it.

So when Jerry Colonna invited me to join a private retreat in the Rocky Mountains of Colorado—not as a journalist, but as a participant—I immediately said yes. He held these retreats regularly for founders he coached, and I didn't know much about them, other than the fact that he'd told me people who'd attended seemed to come out of the woods after the weekend with a sense of clarity. I didn't ask for more details. I just said yes.

With only a couple of days left until I had to let CNN know my decision, I thought a remote cabin, surrounded by roaming herds of bison, would be a much-needed escape—until I arrived and realized I was surrounded by the many founders I'd interviewed.

You've got to be kidding, I thought, with a smile that I hoped would mask my horror as I waved at one founder whose service powered a third of the internet. The camera that followed me when I spoke to these folks was gone. The mic. The makeup. I felt raw. Completely exposed. The only things I had with me were my crumpled notes I'd written ahead of time, part of the homework we'd been asked to do.

Does my life have meaning? I'd written in messy blue letters. *It's important to strip away the screens and filters and assumptions that have been limiting your vision and your action. Start again. Stop. Do over.*

If they ask me to read this out loud, I'm feeding myself to the bison.

When everyone had arrived, Jerry gathered the twenty of us in the rustic lodge filled with wooden beams and cozy leather couches, where we were asked to open up to each other, entirely off the record. As snow began to fall, I wished I'd brought twelve extra layers for warmth and camouflage.

I listened as entrepreneurs, many of whom were powerhouse female leaders, shared their deepest hopes and darkest fears. These were things they'd never told me before, would never say in front of a camera. It was liberating and honest. I found common ground with many of the people in the room. I wasn't interested in their stories to cover them; rather, it was helpful to hear them in order to better understand myself, and slowly start building the springboard to jump. And then it was my turn. I was embarrassed to be on the other side of the equation, sharing how my picture-perfect job didn't fulfill me, how I wanted more. But as I shared my own story, one I'd never truly articulated before, I settled into the discomfort of opening up.

The next day, Jerry instructed us to go and talk to trees. It was weird and woo-woo in a way that normally would have sent me running, but if the CEO of a billion-dollar business could talk to a tree, so could I.

I walked into the middle of the peaceful woods as the snow fell around me. Away from the traffic, the sirens, the lights, the push notifications, the world seemed to stop. Eventually, I found myself at a large tree, its limbs stretching toward me. I looked around, still aware that people I'd known throughout my career were nearby. *They're also talking to trees*, I reminded myself, wondering by how much the net worth of this small patch of land had increased due to tree-talking founders. I caught myself engaging my defense mechanism, and willed the jokes to disappear. I had to start taking myself seriously, even if it meant talking to a tree.

I stood, completely still, in front of an old pine in the Colorado wilderness. The silence was engulfing, almost deafening.

Then the weight of everything hit me, and I cried, tears streaming down my cheeks.

I was back in Atlanta, piecing together bits and pieces of my childhood. The stories I'd used as crutches—the car pulling out of the driveway, my teal JanSport abandoned on the couch of our suburban den—burst open, and for the first time in a long, long time, I allowed myself to feel. My broken family, the throbbing pain of abandonment, my inability to fit in, my ability to draw outside the lines. I experienced the thrill of breaking news, my fear of commitment, the love and then the grief that followed Mike and Ethan. The loneliness of living on planes, the protection of the camera lens, always facing out, celebrating the strength in others instead of in myself.

I thought of Daniel and Deb and Erica, the career moments of the last decade, the big stories I'd waded through that would shape culture and society. I thought of *Mostly Human*, and the fact that refusing to fit inside the lines had forced me to create my own paradigm. To create something out of nothing. To use storytelling as a means of education and empowerment and human connection.

I started to thaw. I was releasing my feelings—fear of abandonment and disappointment, pain combined with resilience.

I put my hand on the trunk and began to speak. "Some would say you're dying, but I think you are standing taller than ever. You have lived."

I could sense another transition coming. I'd always guarded my heart fiercely. And while I'd experienced immense sorrow, I'd never allowed anyone close enough to truly break my heart.

"I've avoided getting my heart broken," I told the worn tree, placing my hand on its rough bark. "Instead, I'm breaking my own heart."

I was unable to move, fixated on the pine. I grieved for the child who was afraid of intimacy. I made a promise, to myself and to the tree, to let go of the fear that sat inside me. The part that said, *You will be disappointed; no one will really love you; you can't have it.*

I'd been standing there for what felt like hours. The snow had started piling up in soft mounds. I wiped away my tears, realizing that I couldn't see the trail.

You'll find your way back, I promised myself. As I waded through the snow, fueled by a sense of urgency, I looked back for one last glimpse of the tree. It was barely visible, but I could see my footsteps leading away from it. I turned around to head back to the cabin, and for the first time in my life, I heard the pure sound of snow falling.

That evening, I found Jerry. I told him about my struggles, with myself and with CNN. I told him about my life's goal of finding Boardwalk. I asked him if I would always be searching for a better relationship, more meaningful work.

His face softened. He adjusted his glasses and let the silence sit. Normally, I would have tried to fill it, but instead I looked outside, where the snow was just beginning to fall again, where stars reflected on the silver mountainside.

"Don't you get it, Laurie?" he said, pulling me back from my thoughts. "Boardwalk is within."

By the time I boarded a plane back to New York, I knew what I had to do.

It was November 26, 2018, and god, was I uncomfortable. Wearing a red dress, just like the dancing emoji girl I saw as my spirit creature, I sat on the DUMBO promenade in Brooklyn. I was two coffee cups of liquid courage deep, watching as children raced to the carousel, where they wrapped their arms around magnificent horses, bobbing up and down against the tall buildings that framed New York. It was the scene of a postcard I could have sent to my younger self—the one who sat in the Esperanto Cafe, writing her own future:

Dear Laurie, you'll manifest your picture-perfect New York dream, and then you'll sit across the water and feel like a stranger to it all.

I warmed my hands against the thin to-go cup and tried to take it all in. Today was the day I would leave CNN.

My meeting with Jeff was in two hours. I took a final sip of coffee and ran through the game plan once again. I'd begin with updates on *Human Code,* as well as on the documentary I'd been planning about Facebook's fifteenth anniversary. I'd butter him up with good news, and then slide into my resignation. It was a solid plan, and yet I still fought the urge to throw up.

As I watched the tugboats glide by, I looked at my city with a worn perspective, and thought about the last decade and the young woman who'd moved to 126 St. Marks with the goal of becoming a journalist. I was proud of my last ten years, and also couldn't help but look at the destruction left behind after a thrilling decade: the algorithms coded by baby-faced entrepreneurs who'd sat in front of me during their companies' earliest iterations, promising to change the world for the better. But our world didn't look better. It was more polarized. It felt like we were all shouting at each other, and no one was listening. And then there was the shifting culture of the newsroom that had become so intricately interwoven with my identity, I could barely envision my world without it. But I knew I had to. I could feel the cool concrete against my legs, as I sat back and thought about the women who'd helped raise me in the newsroom, who'd fought for me behind the scenes. So many of them were gone. Over the years, my mentors in positions of leadership had left, some so they could advance, others because they'd been pushed out, were fed up, or like me, were exhausted.

I wrestled with the fact that some of the people who helped me were the same ones who were now holding me back. It was another gray area. But what was clear was that there were opportunities that seemed to dissipate, battles Erica and I, and many other women in similar positions, kept losing, and wars we weren't sure we wanted to fight anymore. Maybe I couldn't exactly put my finger on it, but looking behind me, it all was there, piled up after a decade at the network, marked with those five words I'd come to know well: *death by a thousand cuts.*

It was hard to grapple with because my story was one of triumph. I had succeeded. I'd risen. Much of that was due to many of the strong-willed journalists and executives, some still at the company, who'd invested in me throughout my career. Yet, no matter which direction I turned, I was hitting a wall. I worked in a culture where the executive suite didn't look like me. And the environment I worked in was beginning to feel the same way I imagined it felt to the women who'd helped bolster me before they'd left. The higher I got, the less of a place I felt I had. It was time for me to own what I did—and not rely on other people to tell me when and how my voice could be used.

Looking out at the serenity in front of me, the children scrambling through Brooklyn Bridge Park, I honored the executives, the producers, the women and men at the company who'd helped bring me to this place, and promised myself I'd build the world I wanted. I would create my own environment where we could all move up. It wouldn't be easy. And I had no idea how to do it. But I couldn't fit in their box anymore.

I wasn't aggressive or ungrateful. I was ambitious, and I didn't want to apologize for it. I sat with complicated gratitude for a company that raised me, and the cuts that wouldn't kill me but would help me give birth to something new. Something that was mine.

Hold on to this feeling, I thought. Stretching my legs, I stood up to walk away.

"Fifty-Eighth between Eighth and Ninth," I told the Uber driver. It was like saying my name, I'd said it so many times over the last decade. *CNN, Columbus Circle.* My second home.

As we passed by the redbrick warehouse buildings and artisanal coffee shops that filled DUMBO, I called my dad with the update. We spoke on the phone regularly, and I leaned on him for advice.

"I can't believe I'm doing this." The more I said it out loud, the more real it became.

I went on, asking the same question I seemed to be asking myself in all aspects of adulting. I looked out at Manhattan, at the boats gliding along the East River.

"How do you leave something that's not terrible, but your heart tells you isn't right?"

I was grateful that I could now ask my father about matters of the heart. We had overcome our skewed version of intimacy and had developed a genuine relationship. My father had become a confidant as I'd struggled with the decision to stay or go, and was one of the first to support my decision to leave my job.

"To live your dream, you have to put your feet on the ground at some point," my dad said to me, before we hung up.

Boots on the ground. It certainly wasn't easy.

"Let me ask you a question." My driver, an elderly Black man wearing a blue jacket, peered at me through the front mirror. He'd clearly caught bits and pieces of my anxious conversation with my father.

"Sure. Ask me anything," I replied.

"Do you love yourself?" His voice was deep and slow, and I worried that my life was beginning to resemble a nineties film starring Meg Ryan. He continued: "Because no matter what anyone says, it doesn't matter if you believe in yourself."

Do I believe in myself? It's time to start.

"Thank you," I whispered, refraining from reaching over to hug him. Maybe it was Jerry's tree talk, but I was becoming a bit woo-woo.

I got out of the Uber and made my way into the Time Warner Center, up to the fifth floor. I changed from flats to heels and waited outside Jeff's office until his assistant said he was ready for me.

As planned, I led with the names, the founders I had already interviewed for *Human Code*. Uber's new CEO. Pinterest's CEO. Salesforce's CEO. I allowed a sense of calm to wash over me. I took a breath, ready to dive into my Big News. Then Jeff's phone rang.

He gestured for me to pause. It was his daughter, complaining about the dentist.

"Don't worry," he said, his tone softening, becoming a dad instead of a major TV executive. "You don't have to go to the kids' dentist anymore."

As I sat there listening, I could feel the confidence leaking out of me. *How could I leave this?* I almost couldn't believe I was quitting. Before my last drop of courage drained away, the call ended. Jeff's bright silver-blue eyes glanced from me to the TV screens behind me, and then back again. I would always be competing for his attention.

With every bit of fortitude, I pulled myself up and introduced the topic of my contract, which I knew CNN was interested in renewing. I wanted to tell him that the situation was too hard, too emotional. That I was terrified. That I was running on empty. But instead, I stuttered, "There are just too many stipulations . . ."

He seemed confused. "What stipulations? We can change that. Just give me four to five interviews."

"But," I said, "if I give them to you, I can't use them for something else."

Was I even making sense?

We started going back and forth, negotiating in circles. I became filled with self-doubt, spinning like the carousel in front of the Manhattan Bridge. And then it came out of nowhere.

"Jeff, I want to tell you how lobsters grow."

He looked at me sitting across from him at his desk, and then moved to the couch, where he sat when there was something important to say.

Then I told the president of one of the most powerful media

organizations in the world about a YouTube video of a rabbi talking about how lobsters grow. As I had found myself tormented by the idea of change, of growing out of an idea of what I *should* be, I'd become obsessed with the video—along with nearly 300,000 other people.

"When lobsters grow, they have to break out of their shells. It's uncomfortable. It's awful. It's the worst thing. That's why this is so hard for me to say," I explained.

He looked directly at me.

"You want to fly," he said.

"I do," I replied, embracing the mixed metaphor. "I've been at CNN for ten years. I need to have a blank canvas. No one is doing what I do. I'm the voice of it."

"You want to be Kara Swisher," he responded.

"No," I said. "I want to be Laurie Segall."

I wanted him to understand what I'd been saying—that technology had become humanity; that we needed empathy; that we needed an authentic voice to cover the movement.

"You want to fly," he repeated.

He got it.

I looked at Jeff, knowing this would be one of our last conversations like this.

"Can I be human for a second?" The words tumbled out.

His clear eyes flickered. "Of course."

"Something I've always appreciated is that even if you don't a hundred percent understand what I'm saying, you trust it, and believe in it." I went on: "You believed in me. I've pitched every position I've had at this company, and I don't fit in the lines. I'll never be a traditional reporter." Saying it to Jeff made it real.

"I know you won't," he responded. "That's not you."

We finished the conversations with talk of logistics: how much longer I'd stay. We said we'd stay in touch. I thanked him and walked out of his office, in disbelief at my own courage.

That evening I walked along the West Side and back to my new apartment in Chelsea. I wrote down a quote in the journal I kept next to my bed. *When everything is uncertain, anything is possible.* I looked around my new place. Much smaller, exposed white brick, and a tiny spiral staircase that led to an uneven roof overlooking the West Side. Nothing about it was perfect; there were unpacked boxes everywhere, and general chaos. But I had a feeling that out of the ashes, something genuine would emerge.

The next day, still in disbelief, I was walking two blocks from the office thinking how in a matter of months, this commute would no longer be mine. But my thoughts were interrupted by an urgent text from Deb: *There's a bomb at the office!*

As I turned the corner to the building, everyone poured outside, but with the typical reporter's dilemma: no one wanted to move too far away, since we were quickly becoming the story. Police blocked Fifty-Sixth Street, calling to bystanders to move away. NYPD vehicles and fire trucks filled the familiar blocks.

I made my way to a group of editors. "Where's Ross?" I said to one of the guys, my heart beating fast. I hoped he wasn't still inside.

Like magic he appeared, wearing a gray beanie and a black jacket. I wrapped my arms around him in a giant hug. We took in the sight: streets blocked off with yellow caution tape. Red and blue lights flashing and sirens screaming atop police trucks, warped and reflected back at us in the shiny buildings on the chilly but clear fall day.

"Is it weird to say 'good morning'?" I asked.

"I'll take it." He laughed.

Apparently someone had sent a pipe bomb to the office. It had been found in the mailroom. Thank god, security flagged the suspicious package rather than opening it. I sent a silent prayer to Gary, my favorite security guard. I couldn't imagine what would have happened if the bomb had gone off in the mailroom or on the newsroom floor. Thankfully, we were informed everyone was okay, including all our security guards.

Standing outside the caution tape, I watched the explosive device being pulled from the building. The media, which had once been a pillar of democracy, had become a magnet for disdain. Trust had been upended by the term "fake news," and opinions and misinformation had replaced facts. My eyes fixed on a large bomb-containment truck, and I pulled out my iPhone to record it.

We would later learn that the plot to blow CNN to smithereens had begun on Twitter and Facebook in the form of tweets and posts. The threats had been flagged but hadn't been removed. Hate speech was moving offline as posts online were weaponized into real-world physical violence. We had reached an inflection point, the clicks and swipes transformed into sirens blaring and police shouting, "Get back!"

I looked over at Ross, the man who'd walked me fifty blocks after my father had called to tell me he had a tumor. The man who'd listened to my teary breakup stories and showed me how to piece together bits of tape to create a heartrending story. I felt so grateful for this man who'd helped me get the job that I was now leaving.

"I'm going to start my own company," I told him.

He nodded, without the least bit of surprise.

"Hire me one day, will you?" he said, winking and giving me a big hug.

"You know I could never afford you," I said, with a watery smile.

Those final weeks at CNN went by in a whir. I wrapped my two specials and said goodbye to all my old colleagues. On my last day, I took my final lap around the newsroom, the familiar buzzing and humming of phones ringing in my ears. The lights were almost blinding, the television makeup heavy, but I felt like a weight had been lifted. The halls were full of the ghosts of the stories I'd chased, the interviews I'd booked, and memories of late nights and early mornings piecing together news segments. I had been head-over-heels in love with this place, had poured my young adult years into the side rooms, the hustle.

And then I went back to see Jeff one last time. In my purse, I had a small paperweight that Ethan and I had bought in New Orleans. It was a statue of a crawfish, but it looked kind of like a lobster.

"I want you to know, out of all my breakups, this one was by far the hardest for me," I said to him, referring to my imminent departure from CNN. "Which, I think, means that it meant the most."

I handed him the paperweight.

He smiled and put it on his desk, next to his framed photos.

Days later, he would sign my goodbye card, *To my favorite lobster.*

Once again, it was time for me to write my own job description.

Epilogue

DOT DOT DOT

I broke my own heart when I left CNN, and it was the best thing I could have done. It was my lobster moment. I'd spent my career covering people who went up against the "shoulds"—the way they should live, the way they should be—but still, I had no idea how hard it would be to walk away from the job I *should* have wanted, from the relationship I *should* have been in, from the life I could have lived. Rejecting all of that took courage, and god, was it hard.

A month after my departure in March 2019, Dennis Crowley asked me to interview him onstage at SXSW, in honor of Foursquare's ten-year anniversary. I brought the old notebook I'd used when I interviewed him for the first time, a decade before. Scrawled in the notebook were the words "ups and downs" and "perseverance"—which seemed to define us both.

Back then, he was a party guy who appeared in posters lining subway stations; back then, I was a production assistant at CNN, pretending to be a producer. Now he was a father of two. He'd stepped down as CEO of Foursquare to serve as executive chairman. I was striking out on my own, without CNN's protective shell. We both knew how far we'd come from our booze-fueled nights during the boom times.

As I sat next to him, a month after leaving my job, I remembered the New York tech scene in those early years—when the days were long and full of ambition and opportunity, and when deals were made over beers in the evenings and multimillion-dollar acquisitions flooded twentysomethings with early riches. I remembered the East Village and my walk-up on St. Marks, and that sliver in time when downtown New York had been marked with Foursquare stickers—after hours of coding and too many beers at Tom and Jerry's, drunk engineers would tag buildings with those stickers, proudly implying, *We were here. We're a part of something.*

But now, all of us had grown up. We were *there*, but I got the sense that many of us who were a part of that "something" were ready for another chapter.

As we closed our talk, I asked Dennis to explain the biggest difference between the person I'd interviewed ten years before and the one I sat onstage with at that moment.

"I went through so many years of my life where I was like, I can't get married until Foursquare is over. I can't have kids until Foursquare is over," he said. "Halfway through, I was like, fuck it. I'm going to try to have a life outside of work. And [I] got married. And have these two awesome kids. And it's, like, the best thing, better than just being Johnny Startup guy." He paused, and then added, "Thank you for making me say that."

"A good way to end a tech thing is to end on humanity," I responded. And I meant it.

• • •

For years, I'd fostered the dream of bringing humanity back to tech. I had a collection of notes and napkins with ideas and bullet points. Finally, it was time to make good on my ambition.

I teamed up with a dear friend, Derek Dodge, whom I'd worked alongside at CNN early in our careers, and we started scheming. It reminded me of my initial days with Deb and Erica. Derek proved to be the perfect calm to my chaos, and over cups of coffee, early mornings, and late nights, we hustled.

We relied on the network I'd built over the years and I started talking to the angel investors whom I'd gotten to know. We covered the walls of our apartments with Post-it Notes filled with show ideas, themes, and company ambitions. It all worked toward our founding mission: to find authenticity in a filtered world and look at complicated issues with nuance and humanity. We thought about what a modern-day media company could look like and promised ourselves we'd build one that quite literally didn't fit in the box of traditional media and television. Our goal was to create shows, books, and films, but also tap into new forms of media, ones that people had yet to start talking about.

We launched Dot Dot Dot in December 2019, and around Christmas, we opened a tiny office above a cannoli stand in Little Italy. But just a few months later, an invisible disease ravaged New York City, and we scattered, leaving our office empty except for monitors and Post-its.

As ambulance sirens became the soundtrack to our lives and isolation set in, in an upside-down world where we could harm our neighbors by merely breathing on them, our concept of seeking humanity through the lens of technology had never felt so visceral and important. Now we were all connected through a camera lens. We Zoomed, liked, scrolled, texted, and clicked; people danced on TikTok; and Instagram was populated with bread-baking and #gratitude. Although, truthfully, it was hard to gauge how any of us were *actually* doing. We craved human touch, and Facebook had become a digital graveyard, as people tagged their loved ones who'd been taken by the virus. There were beautiful tributes and extraordinary acts of humanity everywhere. But the common filter was pain.

We built Dot Dot Dot against the backdrop of a pandemic that would force the world to become even more reliant on and intertwined with technology. The idea of searching for connection took on a whole new meaning—and urgency. Our mission became even more important.

We did our best to create meaningful content to help people feel less alone. From our apartments, we launched a content studio, creating concepts for shows like *Mostly Human* that we believed could move the needle. We produced a podcast called *First Contact,* devoting our storytelling to confronting ethical issues in tech. We struck deals with Silicon Valley founders I'd known for years to create content on new platforms that emerged as a result of the pandemic.

I wrote this book, which I'm proud to say is a product of Dot Dot Dot.

We had good days, and we had terrible days. But even those terrible days were put into extraordinary perspective as Covid-19 took our family, friends, and neighbors.

But 2020 ended up being ripe for media disruption. Niche media platforms would later emerge as trusted venues, and traditional media struggled even before entering a post-Trump landscape. We were ahead of our time, and we are just getting started. While I believe we'll always be finding our footing, building Dot Dot Dot has been by far the hardest and most rewarding endeavor of my career.

During one of those particularly hard days, I called Jerry from the prison of my Zoom box, wondering if I'd ever accomplish what I set out to do in the middle of an unprecedented downturn.

The moment I saw his familiar face appear onscreen, I breathed a sigh of relief.

"We will get the vaccine. Work through the constitutional crisis. We're going to deal with the racial reckoning. Resolve the economic collapse. We'll be able to hug each other," he assured me with his comforting voice.

But it was quite the laundry list.

"And, Laurie?" he added, zeroing in on the concern written all over my face, in my tight smile and inability to maintain eye contact. "The company will figure out how to make money."

I lay awake at night fearing I'd have to let employees go in the midst of a pandemic, that I wouldn't accomplish everything I'd set out to do. And that anxiety was just one part of the existential pain we all felt: the fear that spread across the globe, taking our loved ones and destroying lives.

"But it's always a roller coaster," Jerry reminded me, referring to what both of us knew so well about the entrepreneurial journey. "So the question is: What are you going to do about it?"

I couldn't yet answer the question, but I promised myself I would get better acquainted with the ride.

He continued, "Now tell me about Jon."

I told him about an unexpected surprise: finding the relationship without the question mark.

I smiled and thought about Jon, his crystal-blue eyes, his inherent kindness, the fact that he was in front of me for eight years and neither of us looked up to realize it. That had all changed during the surreal summer a global pandemic stripped us down to the basics of what mattered most.

I recalled Jerry's words: *Boardwalk is within.* A year after leaving CNN, I felt like I was finding my internal Boardwalk, an ease knowing I was taking care of my head and my heart. I'd slowed down. I'd taken a risk that felt authentic to the type of life I wanted to live, and as a result my life opened up. Jon was spectacular.

"Remember, Laurie, purpose and love. Those are your anchors," Jerry said. "Everything else is crazy. Hold on to the purpose and stay connected to love and keep growing."

When I signed off, I took a breath and got back to work. I had to lean into the dots, and trust that they'd somehow connect in the future.

As I move into this next phase, as I face down the unforeseeable trials and tribulations, I have the privilege of looking back at everything I've seen over the past decade. Of knowing, firsthand, that success comes with failure. That CEOs are people. That bubbles need to burst. That Boardwalk is out there, spectacular is out there—but finding it starts within.

Yes, tech has put us in a moment that's very "dot dot dot . . ."—but isn't that the point? All of the uncertainty creates tremendous opportunity for us to show up—as Jerry says—with fierce bravery to be who we truly are. Finding authenticity in this increasingly filtered world is hard. I work on it every day, but it's what challenges me to keep searching—for the next big thing; the million-dollar piece of the puzzle; the relationship without the question mark; the tech that's going to change the world; the life I believe we're all capable of living.

As we try to navigate this new era, it's my belief that the power that brought us to this moment is the same power that will get us through a complicated phase in history. In my experience, it's that combination of discomfort and resilience, of trusting our internal algorithms despite all the external disruption, that creates the best entrepreneurs, the best relationships, and more important, the most fulfilling lives, filled with purpose, love, and growth.

So, here's to a future full of lobster moments, cracking open life, and developing a new shell that fits. It's something only you can determine. Not ones or zeroes, not algorithms. Technology is another layer of our skin. It's time to start feeling again.

Acknowledgments

It has been a lifelong dream to write a book, and to write my own story is a gift. Putting this book out there has felt like walking down the street naked. It's exposing, and the right kind of terrifying. It's also taken a village to get me here.

Thank you to my agent, Becky Sweren, for giving me the courage to share my story, and for helping to guide me through this process. You invested in me. I still remember our earliest conversations in an office full of books and your encouragement (and faith) that one day I'd meet a deadline. You saw me when I was ready to see myself. I will always be grateful.

Thank you to Jessica Sindler and Dey Street for believing in this story, and to Liz Stein for taking it on full force. You challenged me every step of the way. I couldn't have asked for a better team of incredible women.

Thank you to Derek Dodge. You've been my anchor through all of this. You are my sanity, my rock, my business partner, and my best friend. Thank you for always giving me courage and a dose of reality.

Thank you, Mom. You showed me what it is to have an extraordinary heart. I wouldn't be where I'm at without you. Thank you for the love and support you've given me throughout the years. You have always been on the other end of the line. I will always be grateful for you.

Thank you, Dad, for always supporting my dreams, and for not batting an eye when I make moves that don't often make sense, even if it meant attending a purity ball undercover or watching me quit my fancy job. Thank you for creating an environment of love and support.

To my wildly creative and ahead-of-the-times brother, David Segall, thank you for giving me my first tip that technology could be something interesting. I will forever be grateful.

Thank you, Laura Pancucci; while you may not be mentioned in this book, you are every part of it. You are one of the most special characters in my life, and none of this would have been possible without you. May everyone have a Laura Pancucci in their life. You changed mine. You are in every sentence, word, and theme here. Thank you.

To my New York Family: It's pretty incredible to have the best band of misfits in town by my side. Thank you for the emotional support throughout this whole process and, let's be honest, in general. Deb, you are my partner in crime and my constant reminder of empathy and humanity. Dylan and Jodi, you are the friends who throw Passover dinner across the street in the middle of a pandemic so I can have a "shared meal." How did I get so lucky?

Hana—my brilliant, beautiful friend. Our walks are therapy; our friendship has made me a more complete human.

Thank you, Erica Fink, for being one of the ultimate love stories in this book. I'm working on crossing off our final to-do item every day.

Thank you, Jeff Zucker, for continuing to roll the dice on me early in my career and for your entrepreneurial spirit.

Thank you to Sabine Jansen for helping track down valuable moments of the past and for providing organization to chaos.

Thank you, Leslie Wells, for all your help every step of the way, and in getting this book over the finish line. I couldn't have done it without you. Thank you to Raffie Rosenberg for being my right-hand woman at Dot Dot Dot from the start.

Thank you to Carla Klepper for giving me my first writing job at the *Crimson and Gold* and always encouraging me to own my voice.

Thank you to Beverly Osemwenkhaeu and Merrel Daly for helping me feel like my best self.

To Michael Eisenberg, thank you for convincing me to "just press *go*." We wouldn't be here without you.

To Aryeh Bourkoff, thank you for seeing me, Dot Dot Dot, and so much more. Your mentorship and friendship are a gift. Forward to extraordinary.

And last, but my god, not least, thank you to Jon. What a surprise to find you. Thank you for being spectacular.

About the Author

Laurie Segall is an award-winning investigative reporter. She is currently CEO and executive producer of Dot Dot Dot Media, a venture exploring technology through the human lens. Formerly CNN's senior tech correspondent, Segall launched CNN's startup beat and has covered the intersection of technology and culture for more than a decade. During her tenure at CNN, she created numerous in-depth and award-winning investigative series and specials. She has won multiple Gracie and EPPY awards. Segall also served as executive producer and host of CNNgo's first original series, *Mostly Human with Laurie Segall,* a Webby Award honoree. She lives in New York City.